"[Angier] writes with a vividness and enthusiasm that recall Lewis Thomas or Loren Eiseley at their best, but also with an acerbic and highly original wit... [She is] fully *au courant* with the latest developments in biology and evolutionary science... Angier's book is a gold mine." —*New York Times*

"Oh, what a lovely book to read... It was utterly compulsive—I couldn't put it down until I'd finished." —*The New Scientist*

"A hymn in praise of all things great and small... Sheep have the largest animal brains in relation to body size. Dolphins have nasty dispositions. Lovebirds are philanderers. Natalie Angier's *The Beauty of the Beastly* includes a number of such unsettling assertions." —*Publishers Weekly*

"Science writers abound, but gifted writers whose subject is science comprise an exclusive club indeed. Natalie Angier... combines an original and relentlessly curious mind with a witty literary style that lures even the most reluctant reader." —*New Rochelle Standard-Star*

"Pithy, witty, and captivating... One of the best science books of the year." —*Library Journal*

"*The Beauty of the Beastly* proves that Natalie Angier is one of the best guides to what is going on in the forest, the garden, and the laboratory." —*Grand Rapids Press*

"Natalie Angier is one who is constitutionally incapable of writing a boring sentence." —*New York Times Book Review*

"Few writers have ever so vividly described the magnificence of DNA as does Angier, but she's also eloquent on the white-breasted bee-eater birds of Kenya (they hate their in-laws), cheetahs, hyenas, pit vipers, dung beetles ('nature's original recyclers'), cockroaches... No wonder *New York Times* columnist Angier won a Pulitzer Prize." —*San Francisco Examiner and Chronicle*

The
Beauty
of the
Beastly

BOOKS BY
NATALIE ANGIER

Natural Obsessions

The Beauty of the Beastly

The
Beauty
of the
Beastly

NEW VIEWS ON
THE NATURE OF LIFE

Natalie Angier

A Peter Davison Book

HOUGHTON MIFFLIN COMPANY

BOSTON   NEW YORK

Library of Congress Cataloging-in-Publication Data
Angier, Natalie.
The beauty of the beastly : new views on the
nature of life / Natalie Angier.
p.   cm.
"A Peter Davison book."
Selection of revised pieces which originally
appeared in the New York Times.
Includes index.
ISBN 0-395-71816-3   ISBN 0-395-79147-2 (pbk.)
1. Life (Biology)   I. Title.
QH501.A54   1995
574—dc20   94-49675   CIP

Book design by Anne Chalmers

Printed in the United States of America

QUM 10 9 8 7 6 5 4

TO RICK, OF COURSE

CONTENTS

CONTENTS

INTRODUCTION

WHEN I WAS A GIRL, I had a terror of cockroaches that bordered on pathological. This was a particularly inconvenient phobia for a person who lived in a proto-slummy apartment in the Bronx, where roaches had perfected the art of arrogant accommodation among humans who would as soon squash them as see them. My father would squash them with his bare hands. My mother would wield a paper towel or a shoe. My younger brother would stamp them out with whatever tool or appendage was closest to the oily cruds. Not I. No matter how many hundreds of roaches I saw, no matter how repeatedly I reminded myself that they lacked stinging or biting parts and really could not hurt me, I jumped and screeched every time one skittered into view. I could not be in a room with a visible roach and feel at peace, nor could I bear getting close enough to one to kill it. If I opened a cupboard looking for a glass and instead found a roach, I'd go thirsty. Every evening, before venturing into the dark bathroom and switching on the light — an act that would be to roaches as reveille is to sleeping soldiers — I called my brother and begged him to minesweep the room ahead of time. Judging by the enthusiastic sounds of stomping and hooting coming from within, my brother's task was substantial yet not unwelcome. "OK," he'd say, emerging from the room and rubbing his hands together smartly, "I think I got them all."

This is how profound my terror was. I once woke up in the middle of the night and saw a big roach on the edge of my pillow, heading toward me. Such brashness was unheard of: as bustling as the roach population was, it had never bustled into bed with me. I yelped and leaped to my feet, but what was I going to do now? I couldn't very well wake my brother; my parents had little patience for my squeamishness; above all, I could not kill the cockroach on my own.

I decided to cede my space to the enemy. I curled up at the foot of my bed, lying athwart it rather than lengthwise, knees to my chest, head flat on mattress. Uncomfortable and still scared, I nonetheless managed to fall back to sleep. The next morning, I saw that the roach was nothing more than a piece of crayon, which had rolled back and forth on the pillow's indentation and so given me the midnight sense of something small, dark, and alive.

I tell you all this to let you know, in part, why I have named my book *The Beauty of the Beastly*. I hated roaches then, and I still don't like sharing living quarters with them. But in this book, I give them their moment in the sun — whether the photophobic creatures appreciate it or not. I've learned details about cockroach biology that make me want to salute them. Their behavior, the variety of species in the family, the adaptations they have evolved to live with humans or, in most cases, without them — all are part of the great cockroach saga. It is the story of persistence and resistance, of sensitivity and ceaseless change.

Change is indeed the roach's trademark. In the essay "There Is Nothing Like a Roach," I mention the miraculous effectiveness of the pesticide Combat in keeping the urban roach population at bay. Combat does work better than an old-fashioned spritz from a can, but as of this writing, in late 1994, the cockroaches in my Washington, D.C., apartment are starting to get the better of the little black disks. My kitchen is polka-dotted with two dozen Combat parlors, yet some roaches survive. Either the insects have evolved a mechanism for detoxifying the poison or — my belief — they have learned to avoid eating it. After all, I have known house mice clever enough to shun glue traps, leaping like Olympic hurdlers over a series of them in order to reach a bag of ramen noodles on the other side. Clearly these mice had learned something by watching the fate of brethren who'd stepped on the traps. If mice can improve themselves through observation rather than just mutation, why not roaches? And if that elasticity, robustness, and lust for life aren't beautiful, then not much good can be said for evolution, the mother of all invention, the one who stands by the side of the passing biomarathon and cries, "Looking good! Keep it up! Stay alive! Stay *alive.*"

The beauty of the natural world lies in the details, and most of those details are not the stuff of calendar art. I have made it a kind of hobby, almost a mission, to write about organisms that many people find repugnant: spiders, scorpions, parasites, worms, rattlesnakes, dung beetles, hyenas. I have done so out of a perverse preference for subjects that other writers generally have ignored, and because I hope to inspire in readers an appreciation for diversity, for imagination, for the twisted, webbed, infinite possibility of the natural world. Every single story that nature tells is gorgeous. She is the original Scheherazade, always with one more surprise to shake from her sleeve. Of course, I can record only a tiny fraction of those stories, but what I offer represents a larger plea, for all the stories that can be told, for the preservation of nature on her own terms, complete with the golems and creeps and ogres of the world, the roaches, the snakes, the bloodsuckers, the lowlifes, and the brutes.

Beyond writing about the beauty of many stereotypical beasts, I also offer evidence of the beastliness behind our conventional icons of beauty. Beloved dolphins can behave like sailors at Tailhook; orchids advertise faux merchandise; the legendary workers of the field — the birds, the bees, the beavers — in truth spend more time at leisure than the average European; and every creature cheats on its mate, or tries to.

But even this less than exemplary behavior is beautiful in its subtlety. There is always more to be seen behind the first pass, behind the obvious traits that show up during early observations and initially are used to pigeonhole a species or a social system or a gender. I love learning of new findings that overturn or at least complicate abiding verities, even when I may have written about those verities in the past. For example, I include in this book a story about female choice, a field of research that has exploded over the past decade or so. The idea is that the female of many species is the choosy one when it comes to picking a mate, and that her pickiness serves a central role in the evolution of many of the more exaggerated properties of the male, like bright feathers or booming voices. That assumption is predicated on the comparatively high cost of reproduction to the female. She's the one who invests the most energy bearing and rearing young, so she is the one with the greater incentive to select her mate carefully. The disparate cost of reproduction was thought to extend even down to the sex cells. A female's egg is big and filled with protein, fats, nutrients, molecular signals to start the embryo growing; a male's sperm is small and efficiently packaged, nothing more than a serving of genes wrapped in a slippery protein bullet. As the old scientific truism has it, eggs are expensive, sperm is cheap. Small wonder males so often seem willing to blow their pocket change at any opportunity.

Yet that split between the sexes turns out to be a bit too neat. Sperm is not so cheap after all; making it, in fact, substantially decreases the life span of experimental animals like flies and worms, and we can only wonder if it doesn't do the same to a few of our favorite higher organisms.

This recent insight by no means diminishes the importance of female choice to the evolution of male appearance and behavior. Females give a lot more of themselves to their young than eggs, after all. Mammalian mothers carry their babies around and offer up the breast; they have much incentive to be finicky about who fathers their children. Yet just knowing that sperm output exacts a substantial toll on its masculine maker puts the dynamics of sexual behavior in a new and more refined light. You see things you may have slighted before. You see the female make her choice, and then you see the male make his — embracing her as his newly beloved, or walking away, as though thinking to himself, This really isn't worth shaving a few minutes off my life.

In fact, if there is any lesson I have learned in my years of following science, it is that nothing is as it seems. Instead, things are as they seem *plus* the details you are just beginning to notice. New truths rarely overturn old ones; they simply add nuanced brushstrokes to the portrait. Dolphins may be mean-spirited at times, slashing at each other's flesh so brutally that they leave behind gouges, but they also engage in playful and tender behavior, jointly reaching decisions about when to travel, when to fish, when to rest. Hyenas sit at the top of the carnivore's pyramid, with all the ferocity that implies. Unlike lions, they consume every last body part of their prey — meat, fur, skull, bones. The moment two sibling hyenas emerge from the womb, they start mauling each other, usually to the death of one. Yet when a hyena is in a good mood — and if it knows and trusts you — it'll plop all two hundred pounds of itself down on your lap like a pet and beg to be scratched behind the ears.

The sins of the anointed saints, the Jekylls beneath the hides — these are the reasons that I find it so much fun to think of the beasts I write about as protagonists, imperfect heroes all, playing out the drama of their circumstances and opportunities. And I anthropomorphize shamelessly. I assume that nonhuman species have personalities, intentions, emotions, awareness, even dreams and wishes. I do so for the sake of storytelling, and because the continuity of life on a genetic and morphological scale suggests a significant degree of fraternity among the creatures of the earth. Recently I saw in a natural history museum an exhibit of the skeletons of many species: horses, alligators, monkeys, dogs, mice, birds, dolphins, humans. The display made plain how often nature recycles her best inventions, how the limb bones articulate with the shoulders and hips in the same way whether the animal is a quadruped or biped, hoofer or flyer; how the ribs arc out from the spinal column in parallel parabolas; how the thigh is built of one thick bone, the calf of two slender bones; how we all have finger bones, even though the fingers may end up subsumed by flippers. We really are the same under the skin.

Seeing the similarities in skeletal engineering, I felt the usual mixture of contradictory emotions, of ego deflation and goofy communalism. Here I was, just another of nature's prefab animal kits, the standard-issue parts glued and stapled as though in obedience to a numbered diagram, with scant innovation in arrangement beyond a slight widening of the pelvic bone.

At the same time, here I was, blessed with a design that had passed the test of a dozen geological epochs, evidence that life has really gotten the hang of it, of building a mobile body that is strong yet light, supple and enduring, a body that can spin, soar, leap, dig, climb, flee, swing — that can embody life. I felt the beauty of the way that every beast, myself included, is born to move, to solve problems, to make the best of earth and gravity. I thought that bodies built of such analogous components must surely house natures built of similar sensations and inclinations: fear, joy, curiosity, boredom, friendliness, antipathy.

This is not to say that all animals react the same to the same events. (Obviously not. I may run away from roaches, but my cats will run fearlessly, even gleefully, toward them.) But I take for granted that other species are very much aware of themselves and their surroundings — that they have their own versions of consciousness. A spider consciousness. A cardinal grosbeak consciousness. This seems to me an act of courtesy — and an admission of ignorance. We don't know what's in another creature's mind, so why assume it's a blank? Why assume the animal is a programmed robot or a dumb brute, when it seems to be acting with all the neurotic uncertainty you'd expect of any individual that's been thrown willy-nilly into the thick of life? Thus it was with real delight that I learned there is a vigorous debate among naturalists on the propriety of anthropomorphism (Chapter 27).

The traditionalists insist that it is sloppy science and researchers should work tirelessly to keep their emotional objectivity; the iconoclasts argue that you can never understand another being unless you're willing to empathize with it. I come down roundly on the side of the anthropomorphists, although, not being a scientist and thus not required to back up my woolly notions with data, I'll go farther than most, to the point where I'll anthropomorphize plants. We are learning too much about the complexities of plant defense and communication systems to dismiss the possibility that plants, too, have a grasp of themselves and their surroundings.

I also anthropomorphize molecules. Yes, proteins, nucleic acids, steroid hormones: they too are characters in little plays. They move, they spin, they embrace, they succeed or fail. I call the section about the molecular underpinnings of life "Dancing," because that's how I see in my mind what cannot be seen. But there are other ways of imagining the realms of the submicroscopic. Years ago I asked a protein crystallographer, who uses high-energy X-ray beams to map the atomic structure of proteins, what proteins would look like if blown up to ordinary dimensions. He thought for a moment and then said, "Distorted Nerf balls." "Fantastic!" I cried, and his image has stayed with me. The laborers of our body cells, the tens of thousands of proteins that perform core tasks regardless of whether we're aware or indifferent, are nothing more than colorful, squishable toys.

In delving into the science of molecular biology, I'll do anything to come up with similes or metaphors. I do it for myself, to make the abstract concrete, and I do it in writing to keep the plot going. There's no denying that molecular science can be beastly in its difficulty; that is why people so often ignore it. But revolutions should not be ignored, and molecular biology is undergoing a revolution right now. As a science writer, I have sought to understand both the very large — the evolutionary processes that give us life as we see it today — and the very small, the micro-city that is the cell. Here is where science is making the most phenomenal progress, for the simple reason that progress is possible in molecular biology. The tools exist, and, unlike most evolutionary questions, the questions can be broken down into parts that can be analyzed in some meaningful, reproducible fashion. Scientists generally go after problems that can be solved rather than problems that strike their fancy. So while we might wish that scientists were less reductionist, more holistic in their approach to understanding the nature of life, we must be sympathetic with what they are trying to do, which is to parse nature into knowability. Nothing lends itself to parsing more gracefully than the components of the cell.

In the section on molecular biology, beyond conveying specific stories about specific molecules, I try to give an impressionistic feel for what the issues are, the overarching concepts. With the Human Genome Project having been a public spectacle for some years now, DNA and genes have dominated both scientific and popular conceptions of how things work, how life comes to look as it does, why we think and feel and behave as we do. If only we could figure out the entire genetic code of a human being, the argument goes, we would have "it," the great it of selfhood, the recipe for a human being. We would have a chemical understanding of why one person retreats shyly from company while another cannot stand a moment alone; why one can play the violin so well while another has all the musicianship of a leaf blower; or why one person is gay, a second straight, a third a fetishist of women's footwear. Alternatively, somebody will cry, "Don't forget the environment! Don't forget the nurture half of the nature-nurture equation." As though the proclamation of a dialectic gets us closer to solving the enormous conundrum of how life assumes its form and patina. As though it means anything to say that 60 percent of intelligence is hereditary, 40 percent environmentally determined (or vice versa; you choose the breakdown, for many numbers have been plugged into the two slots over the years). What, after all, counts as hereditary, and what environmental? People conventionally think of "environmental influences" as the ways your parents treat you as an infant, or the sorts of television shows you watch as an impressionable preschooler. But at this point, scientists believe the environment encompasses things that happen to you even before you're born, in the environment of your mother's uterus. Thus, if a pregnant woman is under stress severe enough to change her hormonal balance, and if the change is demonstrated to have an impact on her baby, the effect would be called environmental. Similarly, should prenatal viral infection be shown to cause schizophrenia — a possibility now under investigation — that too would rank as an environmental rather than an inherited cause of the disease.

However, what if events in the womb influence the fetal genes themselves? What if hormonal fluxes or other chemical changes in the uterus were to affect the expression of genes at crucial junctures of development, impelling some genes to turn on, others to turn off? Would the outcome of these alterations be considered the handiwork of the environment or of genetics? We are entering a gray zone of biology here, where nature and nurture are so entwined that if you tried uncoupling them, you'd end up with nothing of any meaning. After all, development does not proceed in a void. The chemical sequences that we call genes, the strings of hundreds of As, Ts, Gs, and Cs, cannot reach their potential and make us who we are without being within who we are; they find their purpose in a particular context and, what is important, they are changed by that context. The double helix is a springy, ever-shifting molecule, the lava in the lamp. As its form changes, so too may its function. A region of the helix doubles over, and a gene that once faced forward now tucks inward, where nothing can reach it to shake it awake. An errant hormone attaches itself to the genome like a piece of gum stuck to a movie seat, and that sequence is silenced for minutes, days, months. These are just a couple of examples of how the environment may have its say and start speaking in the language of genes.

I am interested in DNA as a being that moves in space and

time, an organism in its own right. Therefore, most of the stories I tell of molecular biology are those with a tactile, motile spin. I believe the importance of the structure of DNA has been grossly underestimated, so I attempt in a small way to rectify that neglect. I write of DNA bending, the double helix as double pipe cleaners being folded and pleated and looped so that the genes arrayed along their span may be coaxed into speech; the histones, chunky Mickey Mouse–shaped protein groups that cling to the genetic material, condense it to invisibility within the nucleus and lord it over all congress between the DNA and the rest of the cell; and the telomeres, redundant slats of genetic bases found at the tips of chromosomes that tell the cell how old it is and how close to dying. These features of the cell are celebrated far less often than the supposedly indomitable genes, yet they are among the elements that give sense and life to genes. They also serve as a bridge between the text of the speech as it is written in the nucleic acid subunits of the genes, and the sound, fury, busting-loose beauty of the body and brain as it is built. Where DNA exists in the round, that is where nature and nurture commingle.

Beyond describing specific characters — macroscopic or microscopic — I also explore themes that tie those characters together. In the section on adapting, and elsewhere, I look at behaviors that humans and nonhumans alike often indulge in, but that in our current system of world marketeering earn scant attention and less admiration: the fundamental need for play, for joy, for prolonged relaxation. I return repeatedly to stories of sexuality, of courtship rituals and mating strategies among a broad diversity of creatures. Part of my interest in animal sex is prurient curiosity: I like the details of how female meets male, how they circle each other warily, haughtily, venally, the style of their coming together, staying together, drifting apart. Sex stories are inherently interesting, and sometimes that's enough to justify the telling. Yet I often can't help finding a message in a given relationship, an insight into what is and what could be —

what the options are, the solutions to the abiding and imperious problem of perpetuity. An animal in its sexual prime is universal life, behaving as life has from its inception, devoted above all to thrusting more life into the void of the future; and it is the particular, the individual lover, its heat making it new, immediate, its own hunger surely stronger than any hunger before it. A sexual animal is the inviolate individual, the most arrogant of creatures, believing itself for the moment one of the immortals; and the details of its story, of how male meets female, are everything, the only things. They are the creature rousing itself to its fullest potential, and they are the creature at its most complex, its most exposed, its most revealing. They are the animal crying, "Look at me! Watch me perform! I'm alive and I plan to stay alive by strutting my hour across the stage as it has never been strutted before. I am the most gorgeous, the most engaging, the most irresistible specimen you've ever seen. Look at me, I'm here!" So, OK, I admit, I look.

Lest you think me a cheap voyeur, however, I include in the category of sexual behavior parental behavior as well. I believe that the two are strongly linked: how animals tend to their offspring is often a variation on the theme of their mating rituals, governed by similar hormones, as plain or as intricate as the affair that gave rise to the infants in the first place. This may, of course, be a reflection of my femaleness, for females in general cannot but think of sex and motherhood as a package deal; and many of the evolutionary biologists who argue that parental behavior should be considered a sexually selected trait, along with, say, the size of a male's plumage, are women. Yet the truth is that most of the animals I've found interesting enough to write about for reasons having nothing to do with their family life turn out to display noteworthy parental behavior. They offer up their bodies as breast equivalents; they eat feces just to get the nitrogen to protect their young; they strip apart rodents and turn them into predigested stew for the kiddies. Even some insects, alienist machines though they often appear to us, devote years to tending their young. Sometimes the mating rituals that animals engage in hint at their capacity as caring parents, as, for example, when a courting male cichlid fish violently abuses a female to see if she'll fight back — and thus be a warrior when it comes to protecting her babies.

The theme of commonalities, of what we share with the other inhabitants of the planet, extends to our health and well-being. In the section on healing, I focus on medical and health issues, but from an evolutionary and cross-species perspective. As we descend into a state of national obesity, for example, we might benefit from considering how other mammals metabolize and store their body fat. Why does fat fasten on to certain parts of the body so well and to others hardly at all, and why is upper body fat so much more hazardous to our health than fat on the rump and thighs? Why does obesity make a person prone to heart disease and high blood pressure, while a woodchuck can become almost obscenely fat each fall without its arteries paying the price? These are some of the piquant issues that arise when you take an evolutionary slant on a familiar topic, our fatness. The same can be said for menstruation. In Chapter 28, I present a revolutionary view of the purpose of periods that only an evolutionary biologist could have dreamed up.

One section, though, is devoted almost exclusively to human beings, that designated "Creating." The creative impulse may not be confined to us — think of the bower bird or even the dung beetle and its flawless brood ball — but we have taken it to far and away the most extravagant heights; the n of possibility here approaches infinity. That idea struck me recently while thumbing through a magazine and coming across an advertisement — probably for a tourist agency pitching the wonders of Rome — that showed a few postage-sized details of paintings by Raphael, Leonardo, and Michelangelo. Small though they were, the pictures leaped off the page with their magnificence, leaped out of the magazine and its stubborn, two-dimensional dailiness; they were not of the same stuff or species as the graphics or text that surrounded them. The same can be said for a line of Shakespeare or Rilke or Whitman — the roundness of the words, their intonation and texture, the swelling of one phrase, the stillness of the next — none of it sounds or tastes like ordinary language. With their genius, the artists strode far beyond humanness or animalness, or beginnings, middles, and ends — they disengaged themselves from the laws and limits of nature. And so I devote two chapters to the subject of art and genius, one exploring what science can reveal about the neurobiology of greatness, the other considering how the possession of a mortal body and its mortal ailments impinges on an artist's work. Not to limit the subject of creativity to the arts, I include stories about three exceptionally creative scientists, who are not only exploring the world as it is, but inventing it anew through the force of synthesizing intelligence and imagination.

In the final section I return to a subject that knows no species boundaries, the cloak with room to cover us all — death. Here I take a molecular approach, an evolutionary approach, and finally a personal approach to the topic. Emotionally, I hate and dread the thought of death; but intellectually, and from a biological point of view, I acknowledge its rightness, its force and simplicity. Life can be prolonged, but it can never be disengaged from death. Indeed, if you look at the genes that orchestrate the death of a cell — and cell death is the *petite mort* of which our *grande* bodily *mort* is made — they are the same genes that can be subtly altered into agents of immortality. But here's the catch: an immortal cell is a cancer cell. There is no escape, and if any beastliness has a sublime sort of beauty, it is the inviolate facelessness of death.

One last point. Nearly all of the pieces that follow originally appeared in *The New York Times* (with the exception of "Another Stitch in the Quilt"), but they have been substantially revised and personalized for this collection. That doesn't mean I've turned the book into stories about me — "The Dung Beetle and I" — or that I rely overmuch on personal pronouns, unless

I'm there, hands-on, with the beast and want you to know exactly what it feels like to stare an angry three-foot rattlesnake in the eye while stroking its vibrating tail. Nor do I personally condone or believe in all the theories I present, a couple of them being, by the most generous description, highly speculative. But the ideas do reflect my sensibilities and ways of thinking about nature. If you consider them only long enough to snort in derision, at least I'll have kept you amused.

I

LOVING

1

MATING

FOR LIFE?

AH, ROMANCE. Can any sight be as sweet as a pair of mallard ducks gliding gracefully across a pond, male by female, seemingly inseparable? Or, better yet, two trumpeter swans, the legendary symbols of eternal love, each ivory neck one half of a single heart, souls of a feather staying coupled together for life?

Coupled for life — with just a bit of adultery, cuckoldry, and gang rape on the side.

Alas for sentiment and the greeting card industry, it turns out that, in the animal kingdom, there is almost no such thing as monogamy. As a wealth of recent findings makes as clear as a crocodile tear, even creatures long assumed to have faithful tendencies and to need a strong pair bond to rear their young in fact are perfidious brutes.

Biologists traditionally believed, for example, that up to 94 percent of bird species were monogamous, with one mother and one father sharing the burden of raising their chicks. Now, using genetic techniques to determine the paternity of offspring, biologists find that, on average, 30 percent or more of the baby birds in any nest were sired by someone other than the resident male. Indeed, the great challenge these days is to identify a bird species *not* prone to such evident philandering.

Among creatures already known to be polygamous, researchers

find their subjects to be far more craftily faithless than previously believed. Mammals, for example, have never been considered paragons of virtue, yet even here revisionism is in order, and experts are recalculating downward the already pathetic figure of 2 percent to 4 percent that represented, they thought, the number of faithful mammal species. To the astonishment, perhaps disgruntlement, of many traditional animal behaviorists, much of the debauchery is committed by females.

By tracking rabbits, elk, and ground squirrels through the fields, researchers have learned that the females of all three species will copulate with numerous males in a single day, each time expelling the bulk of any partner's semen to make room for the next mating. In so doing, the female can accrue a variety of sperm, thus assuring maximal genetic diversity in her offspring. Most efficiently energetic of all may be the queen bee, who, on her sole outing from her hive, mates with a multitude of accommodating but doomed drones.

For their part, males display marvelous ingenuity in their attempts to counteract female philandering. Among Idaho ground squirrels, a male will remain unerringly by a female's side whenever she is fertile, sometimes chasing her down a hole and sitting on top of it to prevent her from cavorting with his competitors. Other squirrels use a rodent's version of a chastity belt, topping an ejaculation with a rubberlike emission that acts as a plug. Still other strategies result in what are called "sperm wars," battles by males to give *their* sperm the best chance of success when the female clearly is bedding around. In numerous mammals, the last male's sperm is the sperm likeliest to inseminate the female. Hence, several males may engage in an exhausting round robin, as each tries, repeatedly, to be the female's final partner. But how better to thwart one's rivals than with the right tool for the job? The male damselfly, for instance, has a scoop at the end of his penis that can be used before copulating to deftly remove the semen of a previous mate.

These fresh bits of information about mating and the near

universality of infidelity are reshaping biologists' ideas about animal behavior and the dynamics of different animal social systems. The research gives the lie to the dreary stereotype that only males are promiscuous, and that what females want above all is one good male. Instead, many animal social systems very likely developed as much to allow members to cheat selectively as they did to enable animals to divide into happy couples. Most pair bonds might thus be mere marriages of convenience, offering both partners enough stability to raise their young while leaving a bit of slack for the occasional dalliance.

Most of the misconceptions about monogamy and infidelity began in Darwin's day, when he and other naturalists made perfectly reasonable assumptions about mating based on field observations of coupled animals. Nearly all birds form pairs during the breeding season, and biologists believed that the pair bond was necessary for the survival of the young. Without the contributions of both males and females to incubate the eggs and feed and protect the babies, few offspring would make it to the fledgling stage. That demand for stability presumably included monogamy as well.

But as field researchers became more refined in their observations, they began spotting instances in which one member of a supposedly monogamous avian couple would flit off for a *tête-à-tête* with a paramour.

Such sightings inspired biologists to apply DNA fingerprinting and other techniques used in paternity suits to help determine the parentage of chicks. They discovered that between 10 percent and 70 percent of the offspring in a nest did not belong to the male caring for them.

Take the familiar black-capped chickadee of North America. During winter, a flock of chickadees will form a dominance hierarchy in which every bird knows its position relative to its fellows, as well as the ranking of the other birds. Come the spring breeding season, the flock breaks up into pairs, with each pair defending a territorial niche and breeding in it. On occasion, however, a female chickadee mated to a low-ranking male will leave the nest and sneak into the territory of a higher-ranking male nearby. That cheating chickadee ends up with the best of both worlds: a stable mate at home to help rear the young, and the chance to bestow on at least one or two of her offspring the superior genes of a dominant male.

Female barn swallows likewise are finicky about their adulterous encounters. When cheating, a female invariably copulates with a male endowed with a slightly longer and more symmetrical tail than that of her mate; the more sumptuous tail appears to be evidence that the male is resistant to parasites, a characteristic of broad appeal to the female. Not only may she help her young to gain the resistant trait, but, by avoiding infested partners, she limits her own exposure to bloodsucking parasites.

Some females that mate promiscuously gain not so much the best genes as enough genetic diversity to ensure that at least some of their offspring will thrive. A honeybee queen leaves her hive only once, but during that single outing she mates with as many as twenty-five drones patrolling nearby. Tabulating her wantonness is easy: to complete intercourse, the poor drone must explode his genitals onto the queen's body, perishing in the exercise but leaving behind irrefutable evidence of an encounter. While the queen bee does have considerable reproductive demands — that is, sufficient sperm to fertilize about four million eggs — any one of the drones could provide enough sperm to accommodate her, so her profligate behavior seems intended to guarantee genetic diversity in her brood.

But there are evolutionary counterbalances that can keep female cheating in check. Females that actively seek outside affairs risk losing the devotion of their mates. If a male barn swallow observes his mate copulating with a neighbor, he retaliates by reducing his attention to her babies. That is why most adulterous encounters are done quickly and covertly, and why biologists call the events "sneakers."

Of course, males themselves hardly merit any haloes. In an

effort to spread their seed as widely as possible, some go to exquisitely complicated lengths. The older males of the purple martin, the world's biggest swallow, will happily betray their younger counterparts. An older martin will establish his nest, attract a mate, and then quickly reproduce. His straightforward business tended to, he will start singing songs designed to lure a younger male to his neighborhood. That inexperienced yearling moves in and croons a song to entice his own consort, who soon after her arrival is ravished by the elder martin. As a result, a yearling male rarely manages to fertilize more than 30 percent of his mate's eggs, although he is the one who ends up caring for the brood.

In a theme to warm the hearts of aging Hollywood turks, older males often manage to upstage, cuckold, or otherwise humiliate the youthful males around them. Male mallards consistently attempt to force sex on females paired to other males. The female struggles mightily to avoid this copulation by flying away, diving under water, or fighting back — and if her partner is around, he joins in the defense. The only males that manage to overcome all resistance and ravish the female are the most mature and experienced ducks on the block.

But rape is not common in the animal kingdom, and males more often rely on physiology for success. In many cases, natural selection favors males with the most generous ejaculation, resulting in the development of some formidable testicles. The more likely a female is to mate with more than one male, the bigger the species' sperm-producing organs will be. Biologists comparing the dimensions of testes among the higher primates have found that chimpanzees display the largest set relative to their body size. Chimps are the ones that live in coed troops with considerable mating and cross-mating by all. The male with the largest quantity of sperm is likely to swamp the semen of his contenders.

Gorillas are bigger in body but smaller in scrotum, an indication that the great apes' social system does not facilitate sperm wars. Through a combination of genuine ferocity and chest-beating bluster, a dominant silverback male manages to control a harem of females with little interference, or competing sperm, from other males. It does nobody any good to carry around large testicles if one male can monopolize most of the females. Rather than waste energy on growing bigger genitals, subordinate males contrive to overthrow silverbacks so that they have a shot at harem privileges.

Human beings have midsized testicles, good evidence, biologists say, that our species is basically monogamous, but there are no guarantees.

Oh, yes. Humans. It takes an intrepid biologist to apply the new findings about infidelity among animals to the study of people — who obstreperously continue to insist on concepts like free will, consciousness, and unpredictability. As it happens, there are plenty of brave theorists around, and many propose that the human urge to cheat has an evolutionary basis.

Babies need long-term care, which probably led to pair-bonding among humans early in our evolution. But even a happily married man could well be driven to stray from his partner by the urge to slip a few more of his genes into the pool. For her part, a woman might philander to mate with a man who looks to have hardier genes, or at the very least offers her a shank of zebra meat for her favors. Many evolutionary biologists suggest that lapses in monogamy helped spawn male sexual jealousy, bringing about such unpleasant cultural manifestations as female genital disfiguration, foot-binding in China, and other mechanisms by which males have controlled female wandering. And though a woman may have good reason to feel similarly enraged by her spouse's infidelities, she is usually smaller and thus hard put to tie down a man long enough to sheathe him in an iron jockstrap.

But women are not entirely helpless. Evolution has provided them with ways to avoid being too closely monitored by men — most notably through the gift of cryptic ovulation. Unlike that of a rhesus macaque, a woman's butt does not turn bright red when she is fertile. Unlike a moth, she doesn't send a pheromonal broadcast of her status onto the airwaves. Nor does she stand by the window yowling like a cat. So the male can't as easily guard her during the dangerous days of ovulation.

To further confuse men, women have breasts. In great apes, conical breasts are a signal that a female is lactating and thus has low reproductive value. But in humans, a woman's perpetually swollen chest makes it ambiguous to males when she's fertile and when she's lactating, again confounding a man's effort to restrain her reproductive activities. Perhaps this explains why men are so fixated on breasts: they're looking for clues. Too bad they're looking in all the wrong places.

2

THE URGE

TO CUDDLE

IT IS just the potion for a bellicose world. In some cases, it works as an aphrodisiac, inspiring males to seek females more ardently and females to invite their overtures more passionately. It causes the body to swell with arousal and touches off the luscious waves of climax. Afterward, it serves as the classic postcoital cigarette, fostering a feeling of relaxed satisfaction. At other times, it is a promoter of family values, making new mothers more likely to nurture their young and new fathers happier to help out around the nest. Even among those who are neither sex partners nor parents, the compound can induce an overwhelming urge to cuddle.

The magic chemical is oxytocin, a small, potent peptide hormone secreted by the almond-sized pituitary gland at the base of the brain. Oxytocin has long been known as the chemical that stimulates uterine contractions during childbirth and helps the mother's breasts begin secreting milk for her newborn. Now it seems the hormone does far more than serve as a muscle contractor. Scientists find that it is active in both sexes and helps to orchestrate many of life's more pleasurable social and sexual interactions — between male and female, parent and offspring, neighbor and neighbor. It is the satisfactional hormone, nature's way of ushering in joy.

As is so often the case with the most interesting science, much of the work to date has been done on nonhuman species, and nobody can say yet how readily the findings will apply to us. Oxytocin almost surely is a component of human sexuality and happiness, but it is equally likely that the hormonal pathway, as it wends through our bodies and brains, will prove irritatingly complex. Don't expect to see oxytocin tablets supplanting Prozac any time soon.

Nevertheless, oxytocin mania clearly is upon us. Studies of rats, rabbits, field mice, sheep, and other animals show that the hormone acts on many regions of the brain known to participate in sexual and affectionate behavior. The experiments strongly suggest that when the body has been primed for reproduction by sex hormones like estrogen and testosterone, oxytocin is the signal that makes the creature act on the message and scout out a partner. In one study, researchers at Rockefeller University, in New York, discovered that a female mouse given an extra dose of oxytocin during ovulation is 60 to 80 percent more industrious in ensuring that males mount her than is a female not given the shot. The randy female arches her back and displays her hindquarters more fetchingly.

Under other circumstances, oxytocin helps strengthen the bond between parent and offspring. Mother rats treated with oxytocin will pick up and nuzzle their pups more frequently than will female rats without the extra oxytocin; father rats given oxytocin are more likely to build a nest for the pups and guard them zealously. But when male rats are injected with a drug that blocks the activity of oxytocin, they not only neglect to nurture their offspring — they may go so far as to view the newborns as a handy source of food.

Beyond its sexual and reproductive value, oxytocin seems to be a fraternizing compound. Field mice, which tend to crave the companionship of other mice, are naturally responsive to oxytocin. When they are given supplemental doses, the rodents will spend even more time in physical contact, striving to get so close that they are practically crawling beneath one another's fur. But in a closely related mouse that is by nature a solitary creature and shuns its peers except for quickie sexual encounters, the brain has a very different architecture, one that is far less sensitive to oxytocin. For these mice, shots of the hormone fail to work their affiliative magic.

The difference in the two responses may show that what is most important in oxytocin's effect is not so much the total amount of the hormone circulating through the blood as it is the brain's receptivity to the substance. And that sensitivity is dictated by a complex set of factors, like sex hormones and other proteins in the body.

Thus, oxytocin may elicit any number of responses or no response at all, depending on the body's general biochemical balance of the moment. Scientists have yet to figure out which hormonal signals indicate that oxytocin will spark a given response. They believe that, while unsocial species are not likely to possess the necessary equipment to respond to a hormonal call to friendship, in naturally social species it is oxytocin that brings everyone together.

Among humans, the link between oxytocin and behavior is considerably more elusive but nonetheless provocative. In two experiments examining the role of the hormone in men, researchers determined that in the moments preceding orgasm and during ejaculation itself, the level of oxytocin in the blood is three to five times higher than it is otherwise. It could be that oxytocin is needed to start the contractions of orgasm; it could be that the hormone is crucial to the intense feeling of pleasure associated with sexual crescendo; most likely, it is both of the above.

As further evidence of oxytocin's importance in sexual cupidity, normal adults who are given a drug that reduces their sexual desire, thus mimicking a common type of sexual dysfunction, can be restored to their former lustiness with booster shots of oxytocin. But it remains unknown whether oxytocin will help impotent men or frigid women whose troubles are not the result of voluntary consumption of a libido suppressant.

Fresh and surprising though the latest research may be, oxytocin in fact was one of the first neuropeptides, or brain hormones, to be studied in detail; it was discovered in 1903 and its protein composition determined by 1950. It proved to be a small peptide, made up of a mere nine amino acid building blocks, which makes for efficient passage from the brain through to the pituitary and into the bloodstream. Until recently, though, studies of the hormone have been largely confined to its role in uterine contractions and milk production. It was viewed in pragmatic, clinical terms and was synthesized in the lab to create the familiar drug pitocin, now used to speed up labor or to encourage the placenta to be released more readily.

Yet hints of oxytocin's complexity abounded. For one thing, scientists knew there were significant levels of the hormone in men, a peculiar finding for something that was supposedly a childbirth hormone. For another, they discovered closely related peptides in creatures as evolutionarily ancient as primitive fish, and such nonmammalian species clearly do not need oxytocin for labor or nursing. Most important, experimental tools developed in the last several years have allowed scientists to analyze precisely the brain's receptors for oxytocin, the proteins on neural cells that clasp the little peptide and permit the brain to respond.

By mapping the distribution of the receptors, researchers have learned that the impact of oxytocin on the brain is broad, and they have begun to get a notion of which regions are especially sensitive to the hormone's signal. Receptors are clustered in the parts of the brain associated with smell, sight, the endocrine system, and the control of ovulation. One region of the brain that is particularly rich in receptors is the ventral medial nucleus, famous for its involvement in a variety of sexual and reproductive behaviors.

But just as all riches in life wax and wane, so does the number of oxytocin receptors in any given region, shifting to the rhythm of sex hormones like estrogen and testosterone, which themselves ebb and flow. In rats, for example, during ovulation, when estrogen levels are highest, the number of receptors in the ventral medial region of the brain soars by 100 percent. With more receptors available, the brain becomes more responsive to the oxytocin circulating through it and, in turn, stimulates the body to behave as it should. In the case of the ovulating rat, the female will suddenly have the overwhelming urge to crouch down in front of a male and display her genitals, the posture known as lordosis, which is a prelude to being mounted.

In the oxytocin-signaling process, timing seems to be everything. Under some circumstances, injections of the hormone in male rats and monkeys produce erections and prompt them to begin mounting females. In other circumstances, a shot of oxytocin will cause a male rat to walk away from a potential partner, not because he is averse to sex, but because the hormone makes him feel already sated. Oxytocin receptors stipple the regions of the brain known to be involved in the postclimax sensation of sexual fulfillment, a condition called satiety.

In fact, the complexity and versatility of oxytocin underscore just how thin the line is between one of life's pleasures and the next, between excitement and satiation, between the urge to cuddle and nurture and the urge to copulate. Mothers have reported feeling sexually stimulated while nursing their young, and it is quite possible that the oxytocin produced to help release milk has the added effect of arousing the mother — especially since the estrogen levels that control oxytocin receptiveness in the brain are sharply elevated during nursing.

Perhaps the clearest example of the link between sexuality and nurturing can be seen in sheep. Mother sheep will normally take care of their lambs only if they spend the first six hours together. If there is a problem during birth and the lamb must be taken away for longer than six hours, it is too late: the ewe will either abandon the lamb or ignore it. But as sheep farmers in New Zealand and Australia have long known, they can encourage a ewe to bond normally with a lamb, even when the two have been separated for days, simply by stimulating the mother sheep's genitals for five minutes. There is a sound scientific basis for the folk remedy. Vaginal stimulation, it seems, releases oxytocin into the blood. And that hormonal pulse is all the ewe needs to commence licking her lamb, bleating softly, allowing the lamb to nurse, and otherwise behaving like a model mother.

3

TELL A TALE
OF IN-LAWS

FOR MANY OF US, a visit with the in-laws ranks as one of
life's little blisters, an experience just slightly more agreeable
than, say, a CAT scan. But we humans have nothing on the
white-breasted bee eaters of Kenya. These birds are such outrageous relations that when a young newlywed female moves into
her husband's territory, ready to lay her eggs and get a family
under way, her parents-in-law will do everything in their power
to wrench the young couple asunder. They do it with such cun-
ning grace and winsome guile that often their son willingly
abandons his bride to move back in with his mother and father,
where he devotes himself to the care and feeding of his parents'
new brood.

Scientists studying two large flocks of African bee eaters discovered that the elder members of the colonies would attempt
to manipulate, exploit, wheedle, and sweet-chirp their younger
kin, all in the hope that the more callow birds would forfeit their
independence and instead choose servitude. The findings reveal
a new twist in the already snarled web of family life among
social animals, and they offer a tart view of the evolution of
cooperative behavior, one of the most compelling questions in
biology.

In a typical bee eater encounter, it is the father who seeks to
woo his grown son back to the parental nest, where the son will

be expected to help gather insects to feed the father's latest clutch. The senior bird does not try to get his way by mauling or bullying his son or daughter-in-law; an adult bee eater is about the size of a thrush, and it is hard for one to push around another. Nor does the patriarch indulge in dominance displays or flaunt his more mature masculinity. Bee eaters are gorgeous birds, with opalescent bellies, emerald backs, blue tails, and shimmering splashes of red and black; males, females, young, and old are all similarly adorned.

Instead, the father becomes a jovial but persistent pest. He visits the newlyweds dozens of times a day and disrupts their efforts at housekeeping. He plops down outside their nest and blocks their re-entry. When the son tries to fatten up his bride in preparation for egg laying, the father nudges in and begs for the food. All the while, the elder bird punctuates his unseemly behavior with the friendly little gestures of bee eater sociability and solidarity: he quivers his tail, chatters his bill, cheeps softly. About 40 percent of the time the son, perhaps with a stifled sigh of resignation, concedes defeat and moves back home to help raise his siblings. The deserted female is left to languish around her own nest with little to do. She may even have already laid a few eggs, but without her mate's assistance, she cannot rear the chicks.

The bee eater's story offers the most spectacular evidence of what Dr. Steven T. Emlen, of Cornell University, calls "the darker side of cooperation," the efforts by some members of highly social animal species to wrest from their relatives a degree of assistance and sacrifice extending far beyond the call of duty.

In many species of birds and a few gregarious mammals like mongooses and wild dogs — societies where parents, grandparents, aunts, nieces, and in-laws all breed and feed together in close quarters — some acts that look like blissful cooperation between kin are actually subtle forms of exploitation.

The younger relative in the transaction is not always a total loser in the arrangement. In the case of the bee eaters, the son, by helping his parents raise his brothers and sisters, indirectly keeps some of his own heritage alive through the many genes he shares with his siblings. Nevertheless, he would fare better from a genetic standpoint were he to raise his own chicks, and he would try to do so if it were not for his nagging elder. Hence, the trick is to identify which environmental and social conditions allow older animals to manipulate the youngsters, and which conditions encourage the subordinate creatures to rebel.

Biologists have long known that many species of birds and mammals engaged in what is called cooperative breeding, where one lucky couple in a group freely reproduced while the other adults on the team relinquished their own fecundity, dedicating themselves to the care and feeding of the principal pair's offspring. Such acts of apparent altruism seemed to defy evolutionary sense, which in its crudest reading calls for a monomaniacal devotion to the propagation of one's own DNA. Investigating these cooperative breeders, biologists discovered that, in almost every case, the martyr adults were close kin of the breeding pair, usually children or siblings. Thus, the sacrificers were obeying at least some of the tenets of Darwinism; although they were not bearing their own babies, they were still working for the good of their bloodline.

On further scrutiny, though, researchers realized that the indirect explanation alone did not suffice, and that additional factors had to come into play to justify an animal's decision not to reproduce. They noticed that the nonbreeding adults often had subplots of their own when they opted to help around a kin's nest. Normally, the helpers were relatively young, and some seemed to view the season they spent working at home as a kind of apprenticeship, where they learned to rear young under the safest possible circumstances. More often, animals became helpers when they could not find nesting areas of their own, either because surrounding territory was too crowded with competing members of their species, or because most potential sites were vulnerable to predators. In such cases, the helpers seemed to be playing a waiting game, assisting their elders and hoping the relatives might die off soon and leave the breeding spot to them.

The hoatzin, a bizarre, slow-moving bird found in Latin America, is another avian species that breeds in cooperative groups. The hoatzin eats leaves, digests them laboriously with a ruminant stomach like a cow's, and, when young, climbs around from one plant to another with unique temporary claws that extend from the tips of its wings. The bird has such specialized nesting and feeding needs that all the most desirable breeding habitats — leafy, well-protected little islands in swamps — bristle with scores of hoatzins, each group a squawking, squabbling clan of relatives. Inexperienced hoatzins have a terrible time trying to break out on their own. By the nature of the hoatzin system, the females have to disperse from their birthplace, and they spend months flying about from one territory to the next, desperately seeking an opening in a foreign tribe.

Among other species, young adults may decide to help rear their kin with the hope that the newborns will become their little acolytes when they at last are ready to propagate. Biologists who have devoted twenty-odd years to observing the Florida scrub jay have determined that birds do well to serve as helpers, particularly when most of the prime nesting spots in the neighborhood are spoken for. As a helper jay ages, it attempts to turn the kin it had mollycoddled into winged soldiers that will help it wage war against another bird and appropriate its home.

But in other cases, helpers obviously are on the short end of their transactions with relatives. An extreme example of frank exploitation occurs among dwarf mongooses, rat-sized African mammals that are distantly related to weasels. They live in abandoned termite mounds in groups of twenty or so close relatives, with one pair of adults doing all the breeding and the rest doing everything else (guarding the den, feeding the young, carrying the little pups around). Most remarkably of all, the subordinate females in the group also serve as wet nurses, lactating to feed the babies of the dominant female. "That's an extraordinary investment of resources," said Dr. Peter Waser, of Purdue University, who studies the mongooses. "Milk is very expensive to produce, and females usually will do it only for their own offspring."

The dominant females wrest such dedication from their kin in one of two ways. In rare instances, young and subordinate females do become pregnant themselves, but their offspring always mysteriously disappear, very likely the victims of infanticide by the alpha couple. And once those bereaved mothers are producing milk, they may as well suckle the young that are around, greatly increasing those pups' likelihood of surviving. In another, more recondite process, subordinate females simply begin lactating just as the pups of their domineering relatives are emerging. That outpouring is particularly impressive because the helping females are often sexually and hormonally repressed in other ways — which is why they have such difficulty getting pregnant. They are maintained in their ahormonal state through constant, low-level stress, as the dominant female perpetually reminds them, through bullying nudges, chemical memos, and other displays, that she is the queen of the clan.

Subordinate females might be expected to rebel against their unjust fate, but they have few options beyond patience. In general, they fail miserably when they attempt to stake out their own territory. Mongooses have many enemies in the African savannah, and in fact their extreme vulnerability is probably the reason they evolved as social animals in the first place: it pays to have a network of kindred around to keep watch for predators.

Over the fifteen years that Dr. Waser and his colleagues have followed the mongooses, they have seen twelve instances of young adults who wandered off to establish their own homes. Of all those bold ventures, only one produced an offspring that managed to reach adulthood. Beyond the safety benefits of staying at home, it seems that the older the females get, the likelier they are to be given the opportunity to breed. Even when the dominant female remains in the mound, still clearly in charge and still the biggest breeder of them all, she permits the older females to have at least a few pups.

Sorry as the underling mongooses' lot may be, it is easy compared with that of the young female bee eater who laid her eggs only to have her mate abandon her in favor of his parents. While the son benefits indirectly by rearing his siblings, the female reaps nothing from his capitulation to his father, and stands a good chance of losing all for the entire breeding season. She may, however, try to get her revenge. On rare occasions, a forsaken female will seek to save the unborn offspring that were sired by her spineless spouse. She will sneak them into the well-tended nest of her parents-in-law and in a roundabout way win her husband's help after all.

4

FEMALE CHOICE:

AN EVE-OLUTIONARY FORCE

WHAT DO females want? Every man who has ever rolled his eyes heavenward in exaggerated bafflement at this enigma might do well to follow the example of evolutionary biologists: stop scratching that little worn spot on the back of your skull and start paying attention to the evidence at hand. In laboratories and field research stations across the United States and abroad, biologists are exploring an evolutionary force that has long been neglected: the effect of female choice on the appearance and performance of their mates.

The new results suggest that many absurd and seemingly irrelevant courtship rituals and displays that males so strenuously engage in serve a crucial purpose in allowing a female to judge the robustness or health of her potential mate before committing herself to the union. Biologists have suspected for years that certain flamboyant features among males, like the peacock's technicolor tail and the bullfrog's booming moonlight sonatas, evolved for no other reason than enabling males to curry favor with females. But many dismissed the role of female choice as a minor influence in evolution of animal traits, compared with the ability to elude predators or defend territory or with warlike competition among males for access to females.

Now the female animal has finally come into her own in the biological arena. Through the graces of improved research tools and more sophisticated evolutionary theories, biologists have more precisely measured the features that entice females to mate with one male rather than another. As a result, the study of courtship rituals in animals has at last evolved beyond the status of a parlor game — an entertaining round of Kiplingesque just-so stories — to a rigorous discipline, based on experimental work in which characteristics of a male are subtly manipulated to see how changes in these traits affect his desirability and batting average.

The work suggests that females of many species give close scrutiny to telltale signs of parasitism or disease. Hence, the males often signal robust health by sporting skin colorations or feather patterns that become accentuated or exaggerated over generations of selection by females. Some female birds and frogs demand of their suitors a performance that pushes the males to their cardiovascular limits as a test of the hardiness of the males' genes.

In other species, especially insects, a female will refuse a male's sexual overtures unless he offers her some nuptial gift, a nutritious packet of protein and nutrients laced with a defensive chemical that she can use to protect herself or her eggs.

The long-term consequences of female choice affect the females' characteristics as well as their mates', for daughters presumably inherit from their mothers a predisposition to favor certain masculine traits over others. Yet the case for female choice should not be overstated, and the whys, wherefores, and even whethers of female taste remain elusive for the great majority of species, including, incidentally, our own (to which many a woman might well respond: What's to choose from?).

Newly animated though the field may be, the study of female choice began at the beginning, that is, with the founder of modern evolutionary theory. Charles Darwin proposed in 1872 that female animals could exert pressure on the evolution of their species in their mating decisions. Biologists, however, long neglected the notion, being mostly male and thus predominantly interested in the behavior of male animals, especially the violent clashes among males during their annual rutting frenzy. The theory of female choice began its comeback in the mid-1970s, when biologists turned away from studies of group behavior and instead focused on the actions and reproductive strategies of individuals in a species. Fleshing out the ideas of Darwin, animal behaviorists proposed that females usually have a larger stake in reproduction than males do. The stake is especially high in female mammals that bear their young and care for the offspring after birth, but even for insects and fish, which invest far less time in rearing young, the amount of energy needed to produce the nutrients, fat, and protein of an egg is greater than that required for generating sperm. As the old dictum had it, eggs are expensive, sperm is cheap. Given their greater investment in reproduction, females presumably have greater incentive than males to seek the best possible mate. Males that want to pass their genetic heirlooms to future generations must either appeal to females or suffer a genetic dead end.

Perhaps the most straightforward work on the fine points of female finickiness has come in species where the female seeks material help from the male in the rearing or protection of the young. In the courtship behavior of a beetle species called *Pyrochroidae*, for example, the male beetle goes through a bizarre ritual of displaying to a potential mate a deep cleft in his forehead. The meaning of that cleavage had long been elusive, but scientists now know that it encloses a tempting sample, a small dose of the chemical cantharidin, familiarly known as Spanish fly. The male obtains his cantharidin by eating the eggs of a blister beetle; then, during courtship, shows the goods to the female. She grabs his head and immediately laps up the chemical offering within the cleft. Apparently impressed with the hors d'oeuvre, she allows the male to mate, at which juncture she receives the real meal. During intercourse, the male transfers to the female a much larger quantity of cantharidin, which she can incorporate into her eggs to protect them against ants and other predators.

In displaying his cleft during foreplay, the male essentially gives the female a little teaser, as though showing her a fat wallet and saying, "There's more in the bank where that came from." Proof of the central importance of cantharidin to *Pyrochroidae* mating came when beetles raised in the laboratory without access to the chemical failed dismally in their attempts to woo females. The most frustrated resorted to rape mounting, but the female beetles were remarkably adept at shaking molesters from their back.

Less obvious than a nuptial gift is what a female seeks when she chooses on the basis of a male's appearance. In one lovely series of experiments, scientists from the University of Bern, in Switzerland, examined the impact of a male's coloration on female choice among the three-spined stickleback, a small fish. The researchers knew that in the breeding season, the male stickleback turns bright red and, on changing color, displays itself before a female in a mating dance of zigs and zags. The biologists also knew that males exposed to parasites turn a dimmer shade of red and remain pale even after they have shaken off their affliction. The question: Would females prefer males that sported the bright red color, signaling current and prior health? To solve that puzzle, they tested female responsiveness to groups of brilliant red males and dimmer, previously parasitized males, first under natural white light, in which the females could see the intensity of the red color, and then under green light, which disguised the relative tones.

The scientists found that when females could distinguish red males from their drearier counterparts, they almost uniformly paired up with the brighter ones, although both groups of suitors performed the zigzag courtship dance with equal zest. But females choosing males beneath a green light arbitrarily picked males of either color.

Like fish, birds are subject to considerable infestation by parasites, a hazard made greater when they settle into nests that offer warmth and cover to numberless avian bloodsuckers. It is not surprising, then, that female birds, too, are preoccupied with parasites. Through experiments with red jungle fowl, the feral relatives of barnyard chickens, biologists have identified the specific ornaments that most attract a hen to a rooster: the comb and the wattle. Hens pay closer attention to the condition of the male's comb and wattle than to any other characteristic, including his size, weight, the aggressiveness of his strutting, and the state of his feathers. The longer the rooster's comb and the brighter his wattle, the more likely the hen is to choose him over a competing male. Her reasoning is impeccable. The comb and the wattle, because they are the rooster's fleshiest features, would be the first body parts to show signs of parasitism and disease — they're the canary in the coal mine, if you will. As it turns out, the state of combs and wattles is also the farmer's measure of the health of a flock.

Beyond resistance to disease, another factor that females seem to find alluring is stamina. Among gray tree frogs, males will sit for days attempting to attract females by repeating a series of trills that they can vary in both length of individual pulses and timing between pulses. In generating their baritonic music, the male frogs consume a huge amount of oxygen and deplete their body's fuel rations. They essentially exercise to the point of exhaustion, as though the females demanded that they push against their physiological limits. And female frogs do like a show of supra-amphibian strength. When given a choice between a sound speaker playing the normal calls of male gray frogs, and a speaker producing synthetic pulses at double the rate that any real frog could manage, the females fling themselves en masse at the source of the fast-trilling songs and do their best to find the unearthly prince hidden within. Why a female wants a partner capable of such aerobic ribiting is not clear. Because a male tree frog contributes nothing to the business of reproduction beyond his genes — no defensive chemicals and no caring for the young — a female selecting a male for stamina presumably hopes to gain from the exchange the probability of begetting vigorous young.

But proving that hardy males sire hardy offspring has been the subject of great contention. Some of the strongest evidence supporting the link comes from a Swedish study of barn swallows, in which the males have tails that are about 20 percent longer than those of the female. To determine whether female choice had anything to do with the extended male plumage, experimenters cut feathers off the tails of some male birds, and glued extra feathers to the tails of others. When permitted to choose between the short-tails and the long-tails, the females invariably selected the more amply endowed males. Again, parasites were at the root of it. Those males with naturally long tails proved to have measurably fewer bloodsucking mites on their bodies than did the short-tails.

To investigate whether the long-tailed birds had a genetic resistance to mites that could be passed along to future chicks, the biologists followed the swallows through several generations and manipulated conditions along the way. They exchanged eggs that had been fertilized by long-tailed males with those spawned by shorter-tailed males, to offset the contributions of environmental factors. After making the switches, they infected the nests of all the birds with the same number of mites. As the young birds grew, the differences in paternity came through: those sired by the long-tailed males had significantly fewer parasites on their bodies than did the babies of short-tailed males. The chicks' resistance to mites had no correlation to which nest they were in or to the number of parasites crawling across their foster parents' feathers; rather, it was determined by the relative parasite load on the natural father, a strong indication that resistance is hereditary.

Why long tails and parasite resistance correspond remains mysterious, but female swallows clearly like their lovers long. About 15 percent of the males in any given flock fail to mate at all, and they are always the birds with the shortest tails.

However, what's true for barn swallows may not be true for anybody other than barn swallows. Many cases of female choice prove to be somewhat arbitrary. A female may pair up with an especially loud male, for example, just because he is the one she hears — or because he's the only one she's met who isn't gay, married, or unemployed.

5

WHAT MAKES A PARENT
PUT UP WITH IT ALL?

PARENTHOOD may be the most natural task in the world, but, considered objectively, the job description matches that of, say, a serf. There's building a nest, struggling through labor, suckling the newborn or fetching it food, cleaning up its messes, beating back predators, and fussing over the infant's every mewl, whinny, and whine. What mad potion would inspire any creature to take on such a job, and to do so with a zest that looks like . . . pleasure? Although it may seem awfully unsettling and counter-Rockwellian to think that one's parents needed any incentive beyond one's cherubic lovableness to keep dumping out those diaper pails, in fact the biochemistry of parenthood is proving to be a carefully choreographed affair. After much fumbling about in the dark, scientists at last are learning which hormonal signals impel males and females to pair up into cooperative units and assume the demands of rearing and protecting their young.

The work has focused largely on rodents, which display a fairly rigid and predictable set of behaviors that can be manipulated and understood. Studies of higher animals like us suggest that the same hormones that shape the dynamics of rodent family life also influence human social behavior, including the mother's responsiveness to her baby, the affectionate bond between male and female, and the capacity of a child to connect with the outside world and form friendships.

The two hormones that appear to be essential to family and other social relationships among mammals are oxytocin and vasopressin, small and structurally similar proteins produced in the brain that divide in their impact along roughly, though not exclusively, sex-specific lines, oxytocin influencing female behavior and vasopressin stimulating monogamous and paternal behavior in males.

The hormones have long been associated with tasks other than controlling behavior. Oxytocin is the hormone that encourages cuddling, birth contractions, and milk production in women; vasopressin contributes to elements of the body's fight-or-flight response to stress, like raising blood pressure. But the power of each hormone turns out to extend far beyond physiology. Nature is, after all, a tightfisted engineer, likely to use the same materials to perform many tasks at once; and the most efficient way to organize a complex series of jobs is by category. If expressing milk and behaving maternally are supposed to occur simultaneously, and if defending one's territory and acting with fatherly regard toward one's offspring are probably linked labors, why not have the same hormone in each case manage the entire suite of activities?

One compelling and dramatic demonstration of the impact of vasopressin, neuroscientists find, is the behavior of the male prairie vole, a small, fuzzy, orange-brown rodent native to the Midwest. Prairie voles are famed for their unusually monogamous and egalitarian ways, the males and females teaming up for life and contributing jointly to pup-rearing duties. Vasopressin is the ingredient responsible for transforming a naïve young male into an affectionate and protective partner and father. That change in behavior begins with the act of intercourse. Immediately after a male has mated with a female, he shows distinct signs of preferring her to other females, not only by cuddling with her, grooming her fur, and exhibiting other familiar marks of affection, but also by attacking strange voles of either sex that approach his turf. Aggression is one way of expressing attachment, and a postcoital prairie vole is a pugnacious fellow to behold.

Researchers have demonstrated vasopressin's role in that change by injecting the voles with a drug to block the hormone. Voles given the treatment maintained their randiness, mating with an available female, but they neither exhibited any great fondness for her afterward, nor did they become aggressive toward strangers. What is more, when male voles were given vasopressin treatments without having mated, they regarded a nearby female as though their relationship had been consummated, preferring her over others and assaulting intruders.

Significantly, vasopressin manipulations of either type had no discernible impact on the behavior of the montane vole, a species that is not monogamous. Male montane voles lack the brain circuitry necessary to respond to the calls for good paternal behavior.

In other studies of prairie voles, neuroscientists at the University of Massachusetts at Amherst have demonstrated that by injecting vasopressin blockers into regions of the brain rich with vasopressin fibers, they could sharply reduce the parental behavior of fathers, inhibiting their tendency to huddle over pups, groom them, and carry them safely into corners.

But vasopressin's effect on behavior is not limited to nuclear family life. Among male rats — creatures that show little taste for pair bonding or caring for young — vasopressin stimulates social memory, allowing males to recognize one another. Rats given vasopressin blockers fail to recognize old acquaintances and begin each encounter with a full-body sniff, a sizing-up normally reserved for newcomers.

The role of vasopressin in human behavior remains in the realm of speculation, but some psychiatric disorders, like autism and a type of schizophrenia, could be the result of depressed vasopressin production. Preliminary studies show that autistic children have abnormally low levels of vasopressin in their bloodstreams, but whether such measurements reflect the activity of the hormone in the brain is unknown. The seamlessness of the vasculature surrounding the brain — a buffer called the blood-brain barrier — keeps the activities of hormones in the skull and body distinct, so a measure of hormone levels in the blood may or may not reflect the hormonal status of central headquarters. Nevertheless, if some way could be developed to get vasopressin across the protective endothelial barrier and into the brain, a vasopressin mimic could, in theory, help autistic patients develop the social attachments they seem so assiduously to shun. Whether vasopressin could ever be used to foster a generation of sensitive, attentive fathers, however, is another and more far-fetched notion altogether.

What vasopressin does for males, oxytocin appears to accomplish for females, encouraging the bloom of social bonds. It goes vasopressin one better on the benignity scale, for while vasopressin elicits both affection and hostile aggression, oxytocin has been associated only with such positive social behaviors as the urge to mate and to care for the young. Moreover, the female brain appears to be wired to respond rapidly and profoundly to a pulse of oxytocin. In recent studies of rats and mice, neuroscientists were startled to discover that the release of the hormone into the brain during lactation almost immediately caused the supportive glial tissue around nerve cells to retract, the first step in the formation of new synaptic connections between one neuron and the next. That the neuronal structure can change so swiftly suggests that the brain is more pliant than anybody had imagined.

Our own brains, too, may be rather more impressionable than we'd like to believe. In studies that she admits are inflammatory in their sociocultural implications, Kerstin Uvnas-Moberg, of the Karolinska Institute in Stockholm, has found that women's scores on personality scales measuring traits like anxiety and aggression change significantly during and just after pregnancy. "Once they become pregnant, they are much calmer and more sensitive to the feelings of other people and to nonverbal communication," she said. "They get higher points in something called social desirability, a willingness to please others."

When they measured women's blood levels of oxytocin before and during pregnancy, she and her colleagues discovered that the sharper the rise in the hormone concentration with pregnancy, the higher the woman's score on the social sensitivity scale. The levels of oxytocin remained elevated during breast feeding, and so did the women's behavioral changes as gauged by the personality tests.

The researchers do not yet know — and they may never know, given the limitations of what can be done with human subjects — whether the association between oxytocin and a pleasant personality is causative or coincidental. Nor is it clear whether the putative hormonal influence on temperament lasts beyond the months of breast feeding. Some of the subjects in the Swedish study seemed to revert to their former levels of anxiety once their milk dried up, while others reported a more permanent change in mood. For some odd reason, they chose calmness over crabbiness, amiability over suspiciousness, even when they no longer had their hormones to blame.

6

DOLPHIN COURTSHIP:

BRUTAL, CUNNING, AND COMPLEX

AS MUCH AS puppies or pandas or even children, dolphins are universally beloved. They seem to cavort and frolic at the least provocation, their mouths are fixed in what looks like a state of perpetual merriment, and their behavior and enormous brains suggest an intelligence approaching that of humans — even, some might argue, surpassing it.

Dolphins are turning out to be exceedingly clever, but not in the loving, utopian-socialist manner that sentimental Flippero-philes may have hoped. Researchers who spent thousands of hours observing the behavior of bottle-nose dolphins off the coast of Australia have discovered that the males form social alliances that are far more sophisticated and devious than any seen in animals other than human beings. In these sleek submarine part-nerships, one team of dolphins will recruit the help of another band of males to gang up against a third group, a sort of multi-tiered battle plan that requires considerable mental calculus.

The purpose of these complex alliances is not exactly sportive. Males collude with their peers in order to steal fertile females from competing bands. And after they succeed in spiriting a female away, the males remain in their tight-knit group and perform a series of feats, at once spectacular and threatening, to guarantee that the female stays in line. Two or three males will surround her, leaping and bellyflopping, swiveling and somersaulting, all in perfect synchrony. Should the female be so unimpressed by the choreography as to attempt to flee, the males will chase after her, bite her, slap her with their fins, or slam into her with their bodies. The scientists call this effort to control females "herding," but they acknowledge that the word does not convey the aggressiveness of the act. As the herding proceeds, the sounds of fin swatting and body bashing rumble the waters, and sometimes the female emerges with deep tooth rakes on her sides.

Although biologists have long been impressed with the intelligence and social complexity of bottle-nose dolphins — the type of porpoise often enlisted for marine mammal shows because they are so responsive to trainers — they were nonetheless surprised by the Machiavellian flavor of the males' stratagems. Many primates, including chimpanzees and baboons, are known to form gangs to attack rival camps, but never before had one group of animals been seen to solicit a second to go after a third. Equally impressive, the multipart alliances among dolphins seemed flexible, shifting from day to day depending on the dolphins' needs, whether one group owed a favor to another, and the dolphins' perceptions of what they could get away with. The creatures seemed to be highly opportunistic, which meant that each animal was always computing who was friend and who was foe.

In an effort to thwart male encroachment, female dolphins likewise formed sophisticated alliances, the sisterhood sometimes chasing after an alliance of males that had stolen one of their friends from the fold. What is more, females seemed to exert choice over the males that sought to herd them, sometimes swimming alongside them in apparent contentment, at other times working furiously to escape, and often succeeding. Considered together, the demands of fluid and expedient social allegiances and counterallegiances could have been a force driving the evolution of intelligence among dolphins.

•   •   •

Lest it seem that a dolphin is little more than a thug with fins and a blowhole, biologists emphasize that it is in general a remarkably good-natured and friendly animal, orders of magnitude more peaceful than a leopard or even a chimpanzee. Most of the thirty species of dolphins and small whales are extremely social, forming into schools of several to hundreds of mammals, which periodically break off into smaller clans and come back together again in what is called a fission-fusion society. Among other things, their sociality appears to help them evade sharks and forage more effectively for fish.

Species like the bottle-nose and the spinner dolphins make most of their decisions by consensus, spending hours dawdling in a protected bay, nuzzling one another, and generating an eerie nautical symphony of squeaks, whistles, barks, twangs, and clicks. The noises rise ever louder until they reach a pitch that apparently indicates the vote is unanimous and it is time to take action — say, to go out and fish. "When they're coordinating their decisions, it's like an orchestra tuning up, and it gets more impassioned and more rhythmic," said Dr. Kenneth Norris, a leader in dolphin research. "Democracy takes time, and they spend hours every day making decisions."

As extraordinary as the music is, dolphins do not possess what can rightly be called a complex language, where one animal can say unequivocally to another, "Let's go fishing." But the vocalizations are not completely random. Each bottle-nose dolphin has, for example, its own call sign — a signature whistle unique to that creature. A whistle is generated internally and sounds more like a radio signal than a human whistle. The mother teaches her calf what its whistle will be by repeating the sound over and over. The calf retains that whistle, squealing it out at times as though declaring its presence. On occasion, one dolphin will imitate the whistle of a companion, in essence calling the friend's name.

But dolphin researchers warn against glorifying dolphins beyond the realms of mammaldom. "Everybody who's done research in the field is tired of dolphin lovers who believe these creatures are floating Hobbits," said one dolphin trainer and scientist. "A dolphin is a healthy social mammal, and it behaves like one, sometimes doing things that we don't find very charming."

Dolphins become conspicuously charmless when they want to mate or to avoid being mated. Female bottle-nose dolphins bear a single calf only once every four or five years, so a fertile female is a prized commodity. Because there is almost no size difference between the sexes, a single female cannot be forced to mate by a lone male. That may be part of the reason that males team into gangs.

One ten-year study covered a network of about three hundred male dolphins off western Australia. The researchers discovered that early in adolescence, a male bottle-nose will form an unshakable alliance with one or two other males. They stick together for years, perhaps a lifetime, swimming, fishing, and playing together, and flaunt their fast friendship by always traveling abreast and surfacing in exact synchrony.

Sometimes that pair or triplet is able to woo a fertile female on its own, although what happens once the males have herded in a female, and whether she goes for one or all of them, is not known: dolphin copulations occur deep under water and are almost impossible to witness. Nor do researchers understand how the males determine that a female is fertile, or at least nearly so, and is thus worth herding. Males do sometimes sniff around a female's genitals, as though trying to smell her receptivity; but because bottle-nose dolphins give birth so rarely, males may attempt to keep a female around even when she is not ovulating, in the hope that she will require their services when the prized moment of estrus arrives.

At other times potential mates are scarce, and male alliances grow testy. That is when pairs or triplets seek to steal females from other groups. They scout out another alliance of lonely bachelors and, through a few deft strokes of their pectoral fins or gentle pecks with their mouths, persuade that pair or triplet to join in the venture.

The pact sealed, the two dolphin gangs then descend on a third group that is herding along a female. They chase and assault the defending team, and, because there are more of them, they usually win and take away the female. Significantly, the victorious joint alliance then splits up, with only one pair or triplet getting the female; the other team apparently helped them strictly as a favor.

That buddy-buddy spirit, however, may be fleeting. Two groups of dolphins that cooperated one week may be adversaries the next, and a pair of males will switch sides to help a second group pilfer the same female they had helped the defending males capture in the first place.

The instability and complications of the mating games may explain why males are so aggressive and demanding toward the females they do manage to capture. Male pairs or triplets guard the female ferociously, jerking their heads at her, charging her, biting her, and leaping and swimming about her in perfect unison, as though turning their bodies into fences. They may swim up under her, their penises extruded and erect but without attempting penetration. Sometimes a male will make a distinctive popping noise at the female, a vocalization that sounds like a fist rapping on hollow wood. The noise probably indicates "Get over here!" for if the female ignores the pop, the male will threaten or attack her.

At some point, the female mates with one or more of the males, and once she gives birth, the alliance loses interest in her. Female dolphins raise their calves as single mothers for four to five years.

The pressure to cooperate and to compete with their fellows may have accelerated the evolution of the dolphin brain. The dolphin has one of the highest ratios of brain size to body mass in the animal kingdom, and such a ratio is often a measure of intelligence. A similar hypothesis has been proposed for the flowering of intelligence in humans, another big-brained species.

Like dolphins, humans evolved in highly social conditions, where kin, friends, and foes are all mingled together, and the resources an individual can afford to share today may become dangerously scarce tomorrow, igniting conflict. In such a setting, few relationships are black or white; it is the capacity to distinguish subtle shades of gray that demands intelligence.

But keep in mind that the dolphin's big brain does not, on its own, rank it as a big thinker. After all, the creature endowed with what may be the largest brain-to-body ratio in nature is none other than the sheep.

7

SKIN DEEP?

BEAUTY is only skin deep. How sweet that old chestnut is, equally comforting to the unbeautiful (who know they have so much beyond physical appearance to offer the world) and the beautiful (who, after years of being pursued for their outer packaging, really do want to be loved for their inner selves).

The only problem with the cliché, if we are to believe some evolutionary biologists, is that it may not be true. In the view of a growing number of researchers who study why animals are attracted to one another, a beautiful face and figure may be alluring not for whimsical aesthetic reasons, but because external beauty is a reasonably reliable indicator of underlying quality. Evidence from species as diverse as zebra finches, scorpion flies, deer, and human beings indicates that creatures appraise the overall worthiness of a potential mate by looking for at least one classic benchmark of beauty: symmetry.

By this theory, the choosier partner in a pair — usually though not always the female — seeks in a suitor the maximum possible balance between the left and right halves of the body. She looks for signs of celestial harmony, checking that the left wing is the same length and shape as the right, for example, or that the lips extend in mirror-image curves from the center of the face. In searching for symmetry, she gains clues to the state of the male's health, the vigor of his immune system, the ability of his genes to withstand the tribulations of the environment.

The new emphasis on symmetry in the choice of mate is one of those annoying developments in evolutionary research that lend oblique validity to ingrained prejudices — in this case, to a fairy-tale view of the world, in which princes and princesses are righteous, strong, and lovely, and bad folk are misshapen and ugly. Biologists emphasize that symmetry is just part of the story of how animals make their choices, and that much remains to be learned about what, in any given species, the possession of a perfectly proportioned body announces to one's peers.

Nevertheless, symmetry does seem to play a role in desirability. When male zebra finches are outfitted with a variety of colored leg bands, the females vastly prefer males with symmetrically banded legs over those sporting bands of different colors on each leg, a sight that apparently has all the appeal of one's date showing up in mismatched socks. Female scorpion flies can detect a male with symmetrical wings either visually or by sniffing the chemical signal — the pheromone — he emits. Given the choice between the pheromone of a male with wings that differ very slightly in length and the cologne of a suitor with matched wings, she will move toward the scent of the even-keeled fly.

Among deer, the male who commands the largest harem not only bears the largest rack of antlers, but also the most symmetrical ones. It is known that a stag who loses a fight to another male — and who is thus likely to lose all or part of his harem to the victorious competitor — will grow an asymmetrical segment on his antler the following year, the sorry obverse of a scarlet letter.

We may talk about the sexiness of a bumpy nose or a crooked smile, but the faces we consider the most sheerly gorgeous — the ones that coolly gaze at us from every fad and fashion ragazine on the stand — are in truth some of the most symmetrical. In one telling experiment, researchers took photographs of male and female college students, put the pictures into computers, and digitalized them for precise measurements. They then gauged the relative symmetry of the faces by putting points on key features: the outer corners of the eyes, inner corners of the eyes, the cheekbones, the outer corners of the lips, the outer edge of each nostril, and the outer points of the jaw. Lines were drawn to connect one dot to its opposite mate, and the midpoint of that line was calculated. On a perfectly symmetrical face, all the midpoints of those lines met and formed one vertical line down the center of the face. Any deviation from that vertical line was a measure of horizontal asymmetry.

The scientists showed the computerized images to other students and asked them to rate the subjects' attractiveness. They learned that, yes, the most symmetrical faces are considered the most appealing. However, the scientists went further, asking their photographed students to fill out questionnaires about when they first lost their virginity and how many sex partners they had had, information that can be considered a somewhat crude estimation of so-called genetic fitness, that is, the likelihood of passing one's genes into the pool.

"It worked like a charm," said the researcher. "Those with greatest facial symmetry lost their virginity earlier, and their number of partners was higher." For men, the possession of a symmetrical face proved better than a witty come-on or even a football letter in winning a steady stream of sex partners.

By an evolutionary reckoning, a symmetrical face and figure demonstrate that the male's central operating systems were all in peak form during important phases of his growth. A well-proportioned body may indicate that the male has an immune system capable of resisting infection by parasites, which are known to cause uneven growth of feathers, wings, fur or bone. Or it may signal a more global robustness, one capable of withstanding such threats to proper development as scarcity of food, extreme temperatures, and ambient toxins.

In theory, females will select a symmetrical male because he can donate superior genes to her offspring or because he is likely to be in good enough shape to help in rearing and protecting their young. The study of symmetry is part of the larger discipline of sexual selection, an intellectually vigorous field that is yielding a host of novel theses on why females opt for one male rather than another. Many outstanding traits found in male animals, from the extravagant plumage of a peacock to the percussive calling of a cricket, are thought to have been shaped over generations by female taste, and the challenge is to understand the sources of that taste. No single explanation will likely suffice. On occasion, females obviously engage in copycat behavior, as one female fish sees which male her rival fancies and then finds herself inexplicably swimming in the preferred male's direction.

Males can be made superappealing by the addition of exotic and frankly unnatural adornments. For example, when a little white feather cap is placed on the head of a male zebra finch, who does not normally have a feather crest of his own, he becomes immoderately popular with females. By contrast, a male given a red feather hat gains no advantage from the haberdashery touch. Neither preference appears to have functional significance. The females could not be choosing the white feather cap as evidence of strong zebra finch genes, because a zebra finch is not supposed to have a hat of any color. Results like these support the so-called sensory exploitation theory of female choice, in which female preferences reflect how an animal's sensory system and brain work — what the brain focuses on and what it ignores in the environment — rather than the outcome of a female's careful appraisal of the male's genetic assets.

Every proposal in the sexual selection trade has its outspoken detractors, and the study of symmetry is no exception. Critics complain that some of the differences in body proportion that scientists now measure are so tiny that they are noticeable only when you put a pair of calipers up to a creature's wings. Is it likely, they ask, that a female scorpion fly, wandering around without the benefit of modern scientific instrumentation, would notice these minute variations between one potential mate and the next? Nor has it been proven that a symmetrical individual possesses especially hardy genes. "People embrace this idea of symmetry because it's something you can go out and measure," said one.

Whatever its precise relevance to animal sexuality, symmetry is an artistically appealing concept, one that painters, sculptors, and architects have been exploring for at least five thousand years, ever since the Egyptians began building their severely, even rigidly symmetrical temples. In his famed paintings *The School of Athens* and *The Dispute over the Sacrament,* for example, the Renaissance master Raphael perfectly counterpoised the people on the left side of his canvas with equal numbers of figures and similar arrangements on the right. Symmetry and proportion were thought to be part of the Lord's plan, an earthly reflection of heavenly perfection.

So, too, does symmetry seem to be part of nature's plan. Most animals have bilaterally symmetrical bodies, with limbs and features mirrored on either side of a central axis, and many flowers are radially symmetrical, all their petals bursting forth in equal arrangements from a central point. Many viruses exhibit an almost mathematical degree of symmetry, as do important structures within the cell that control the cleaving of one cell into two.

Scientists did not begin to appreciate the possible relevance of symmetry to mate assessment until 1990, however. Dr. Anders Moller, an evolutionary biologist at the University of Uppsala, in Sweden, was studying barn swallows, a species in which the males have long tail feathers that grow in a wishbone pattern. He had determined that females like long tails on their males, the longer the better. But in attempting to manipulate the feathers experimentally, he made another discovery: females also like symmetrical tails, with each side of the wishbone configuration the same length and coloration as the other.

Playing around with parameters by adding tail feathers, clipping tail feathers, or painting them different patterns, Dr. Moller found that length and symmetry figured more or less equally in female choosiness. In other words, a long, slightly uneven tail and a short symmetrical tail rated about the same, but when the female was given the choice of a male with a lengthy, balanced tail, there was no contest. Such a well-endowed male was invariably enticing. In other experiments, scientists upended expectations that the biggest male always prevails by showing that female scorpion flies preferred symmetrical males over bigger males with asymmetrical wings.

The work on symmetry meshed with scientific understanding of how disease and pollution affect animal development. For example, fish that swim in polluted waters spawn asymmetrically shaped fry. Why shouldn't any female make assessments of her mate's overall robustness by checking him for symmetry? Seeking evidence that symmetrical animals are indeed sturdier than their slightly misproportioned counterparts, Dr. Moller determined that barn swallows with symmetrical tails are less likely to be infected with parasites than are males with asymmetrical tails. Through test-tube experiments, he also discovered that the immune cells of symmetrically appurtenanced males are comparatively stronger.

Studies of starlings showed that scientists could influence the symmetry of a bird's plumage by giving it more or less food while it was molting. The less food the bird got, the less proportioned the regrowth of its feathers the next year. Hence, the balance of a bird's plumage is a handy and sensitive barometer of a bird's overall health during the present season: the more symmetrical the plumage, the more nourished the male, and presumably the better a provider he will be for his family — rather like somebody who starts the day with a good breakfast probably having more energy in the afternoon than the one who skips the meal.

The obvious benefits of symmetry are not limited to males. The most symmetrical female scorpion flies are the most adept at gathering and hoarding food, fighting off competitors, dominating their peers, and otherwise behaving like members of a ruling caste.

And among humans, for better or worse, female beauty remains the single greatest source of female power, and the woman with the comely, harmonious face will find herself loved, envied, declared one of nature's anointed — at least until the depradations of age begin disassembling the masterpiece, step by asymmetrical step.

8

THE GRAND STRATEGY

OF ORCHIDS

THEY ARE the P. T. Barnums of the flower kingdom, dedicated to the premise that there is a sucker born every minute: a sucker, that is, with wings, a thorax, and an unquenchable thirst for nectar and love. They are the orchids, flowers so flashy of hue and fleshy of petal that they seem thoroughly decadent. And when it comes to their wiles for deceiving and sexually seducing insect pollinators, their decadence would indeed make Oscar Wilde wilt.

The orchid family is the largest of all plant groups, representing upward of thirty thousand species. The flowers are also among the most artfully deceptive, and they have acquired such an extravagant repertory of disguises in color, odor, shape, and overall engineering that for botanists and evolutionary biologists they continue to yield a bounty of surprises. Charles Darwin was so enraptured by orchids that he wrote an entire book about their reproductive strategies.

Yet only now are biologists learning why the flowers are such great pretenders, and precisely how they differ from other plants in such crucial matters as fertility, life span, and position in the ecosystem. The nuances of their biology explain why some organisms reproduce early and often, while others prefer a statelier and more far-sighted approach to ensuring their genetic legacy.

Because most orchids are naturally rare species, their strategies offer profound insights into the tropics and other of the earth's more fragile habitats, where life overbrims with tens of thousands of uncommon beasts and blossoms.

Some scientists are eager to study orchids before the most exotic tropical specimens vanish altogether. Not only are orchids falling prey to the loss of their rain forest habitat, but a renaissance of interest in orchid horticulture is encouraging entrepreneurial plant poachers to strip forests of the most endangered species, which they sell illegally to obsessed collectors not satisfied with standard hothouse varieties. Orchids have "snob appeal," as a horticulturist at the Brooklyn Botanical Garden put it, and snobs delight in having the one, the only, the last and most beauteous of its kind.

Of course, beauty is often in the eye of the beholder. Some orchids look and smell like female bees, presenting irresistible decoys to male bees on the prowl. Others so closely resemble female wasps that the males of the species will molest them time and again, alternately picking up and depositing pollen sacs with each new act of what is called pseudo-copulation. That bit of desperate athleticism may not result in the fathering of any new wasps, but it does help to pass the equivalent of sperm from one orchid flower to the ovary-like structure of a second flower, allowing fertilization to occur.

Another type of orchid has the aroma of rotting meat, coaxing any carrion flies in the neighborhood to come hither. Some orchids mimic the splendor and fragrance of other types of flowers that, in the tradition of floral courtesy, persuade insects to visit them by offering a sup of nectar. But the skinflint orchids do not bother to generate the precious liqueur; instead, they reward any bee foolish enough to fall for the ploy with nothing more than a sticky pat of pollen. Some bees come out of an orchid with such a load of pollen stuck to their backs that they can hardly fly.

The details of the resemblance between orchids and whatever

plant or animal they happen to be aping offer a bird's- or bug's-eye view of how other creatures perceive the world around them and what their sensory capacities may be. Thus, an orchid will evolve a striking pattern if its pollinator focuses on patterns, a chemical if the pollinator is chemo-sensitive, a shape if that's what turns on the desired visitor.

Some orchids do offer a nectar bonus to a pollinator needed to transfer pollen from one plant to another. But the flowers are finicky and concentrate their efforts on beckoning a specific emissary. One type of orchid, the *Angraecum sesquipedale,* a native of Africa and Madagascar, will release a waft of jasminelike perfume in the evening hours to attract a moth that emerges only after dark. The moth happens to have a proboscis, or tongue, that is twelve inches long, just the length needed to reach down into the deep tube of the flower, where its nectar and its pollen can be found.

A similarly magnanimous orchid, found in Central and South America, generates an aromatic oil that the males of a particular bee species need if they are to woo females. After landing on the orchids, the males use little brushes on their front legs to sop up droplets of the precious substance, which they store in their hollow hind legs, releasing it later as an enticement to females. Their close contact with the orchids results in pollen transfer.

But such loving synchrony of purpose is rare, and most orchids are shameless charlatans. Even their name, *orchid,* is something of a ruse. Seeing the wrinkled bulbs found at the base of the orchid stem and believing the orchids originated from these structures, the Greeks named the plant after their word for testicle, the source of human seed. But the bulbs are neither seed containers nor true bulbs, in the fashion of a tulip bulb, which enfolds an entire fetal tulip. Instead, the orchid bulb serves merely as storage for water and nutrients.

Little about orchids is what it seems. Raymond Chandler, in *The Big Sleep,* compared the texture of orchids to the feel of human flesh. Georgia O'Keeffe did not have to use much painterly license to lend her depictions of orchid blooms an erotic feminine air. Many orchids are named after what their flowers resemble: spiders, butterflies, baskets, shoes, peas, and donkeys.

But all orchids have a few details in common: notably, a protruding lip, which tempts insects to land on it as on an airport runway, and an internal column, which contains both the anther cap and the ovaries. Any given flower can either send spermlike pollen to another plant or serve as the recipient for a second orchid's pollen, but few species are capable of self-fertilization. They need a pollinator to pass their genes around. After pollination, a fertile seed pod grows out of the stalk of the flower.

Orchids also are somewhat parasitic. Their seeds are tiny, the size of dust particles, and therefore can carry no protein or nutrients. Once released from the pod and blown to the ground, an orchid seed must gain its nutrition from a fungus growing nearby. Different species of orchid rely on different fungal helpers to germinate. The flower's entire modus operandi seems to be getting something for nothing. Many species are epiphytes, tree dwellers that let their roots dangle slothfully to catch vitamins from bird droppings, rotting leaves, and other material washed down from above by the rain.

But laziness alone does not explain the orchid. Many seem to have about them a touch of self-destructiveness. They can be so mean and deceptive toward their potential pollinators that insects avoid them. Some orchids, for example, use a slingshot system to shoot their pollen capsules at bees that have alighted on their petals; they hurl out the packets with such force that the bees are often knocked long distances. These bees quickly learn to shun the floral snipers.

The pink lady's-slipper orchid, a particularly flamboyant example found in pockets around the United States, has a flower that looks and smells as though it were engorged with nectar, yet not only is it utterly dry inside; it's also a nasty trap. When a bee alights on the lower lip of the lady's-slipper, hoping for a treat, the flower's hinged upper lip closes down, caging the creature inside. The only escape route is through a passageway out the rear. As the bee fights its way to freedom, it must pass by the anther stem, where it inadvertently picks up a cap of pollen. So unpleasant is the encounter that the creature will be unlikely ever to venture near another lady's-slipper. For the orchid, the bee's wariness is dangerous, because it takes two acts of insect gullibility to complete a fertilization: the first to pick up the pollen, the second to smear it on another flower.

In fifteen years of study in a Maryland national forest, one naturalist, who followed the fates of a thousand lady's-slippers, found that only twenty-three managed to be pollinated, presumably by the duncer of the local bee population.

A new theory of orchid strategy explains such apparently counterproductive behavior. It holds that the flowers are nature's quintessential gamblers, willing to bank everything on a potentially enormous payoff. Most flower plants have a high annual rate of fertilization, but when they are fertilized, each produces only one seed or a handful of them. By contrast, although few orchid plants will breed during any given year, when one is fertilized, it hits paydirt. "They work on the lottery system," said Dr. Richard B. Primack, a professor of biology at Boston University. "The chances of any flower being visited are very low, but when the flower is fertilized, it produces tens of thousands or hundreds of thousands of seeds."

A win-or-lose strategy seems to benefit many orchid species. Most of the thirty thousand different orchid species count few members among them. They are naturally rare flowers, living high in trees or in widely dispersed stands on the ground. As a rule, rare species evolve exaggerated, risky, and highly specific reproductive strategies; they are custom-designed to survive in their niches. Many orchids target all their efforts at attracting one type of pollinator. That is why one orchid evolves the shape of a species of female wasp or emits a pungency of interest to one type of fly, or will ensnare one kind of bee and even manage to catch a dimwit twice. Orchids can afford to wait for the perfect pollinator. They are among the longest-lived of all flow-ering plants, and they have very few natural enemies. As a result, far more orchids survive from one year to the next than do most plants.

Suckers may come and suckers may go, but the fakers of the world are built to last.

II

DANCING

9

THE VERY PULSE

OF THE MACHINE

THROUGH A MICROSCOPE, they look like tiny crystal serpents, curving and slithering across the dish with an almost opiated languor, doubling back on themselves as though discovering their tails for the first time, or bumping up against a neighbor clumsily and then slowly recoiling. Beneath their translucent skin the pulsing muscle cells and nerve fibers are clearly visible, a sight so exquisite and so unearthly that it is hard to believe these creatures are common roundworms, found in gardens and compost heaps everywhere. And it is harder still to believe that such slippery squiggles of life are fast changing the face of fundamental biology.

The creature is *Caenorhabditis elegans,* a nematode that measures a millimeter long, feeds on bacteria, and grows from fertilized egg to fecund adult in a mere three days. Keeping pace with the swift development of the organism is the growth of the research field dedicated to probing it in the finest possible detail, cell by cell, molecule by molecule, gene by gene.

Scores of scientists here and abroad are turning to the worm to address the problems that obsess them, from the great mystery of how a single-celled egg can construct itself into a complete, complex animal to the perhaps greater conundrum of the signals that tell a cell it is time to die. Neurobiologists are taking advantage of the nematode's relatively primitive nervous system to study the mesh of synapses, axons, neurons, and sensory organs required for an animal to confront the world.

Of course, the researchers are not interested in worm birth and worm brains per se, although nematode biologists do express affection for their experimental animal. Rather, they believe that what they learn by studying nematodes can be hoisted right up the evolutionary ladder to the study of human beings. The nematode serves as a model organism, a proxy for the rest of us, a specimen that can be manipulated, irradiated, mutated, assortatively mated, plucked apart, scrambled up, over-easied, put back together, sacrificed, and finally *understood* in a way that a human being could never be. Model organisms are the bedrock on which the tower of basic biology is built; without them, we might all have to take turns drinking radioactive dye for the greater good of science.

Looking at the sticky strands of DNA within the worm, scientists have detected genes that strongly resemble human genes known to cause cancer when damaged. Researchers, long stymied by how the human genes work, or by their tendency to turn rotten, can track the behavior of the genes in the worm and thereby gain crucial knowledge about the biochemical mechanism of malignant transformation. Genes that dictate the development of muscle tissue and nerve cells from the primordial cells of an embryo, genes that usher migrant cells to their proper position in the growing spinal cord, genes that incite cells to commit a violent suicide — all these have been detected in roundworms, and all offer a glimpse into the gears, springs, chimes, and whistles of which each life is made. By late 1994, biologists were on the verge of completing a physical map of the worm's genetic information: a collection of isolated chunks of DNA representing all ten thousand or so genes that constitute the creature's hereditary blueprint. The physical map will be the first of its kind devised for any organism more complicated than yeast — a single-celled being — and nematode biologists predict, with the swaggering confidence of true believers, that it will be a magnificent research tool. Geneticists are also beginning to sequence, or spell out, every one of the hundred million chemical bases contained within and between the worm's ten thousand genes. The project, estimated to cost $50 million, is a sideline attraction of the far more difficult international effort to map and sequence the three billion bases of human DNA. While the Human Genome Project remains a subject of rancorous debate and transcontinental rivalry, and hence is proceeding by fits, starts, and stops, worm aficionados hope their sequencing task will be completed by the end of the century, long before the sequence is determined for us or any other higher animal.

The genetic sequencing information will add to an already significant foundation of knowledge about nematodes. Although scientists can only guess at the number of cells in other standard experimental animals like fruit flies and mice, they know precisely how many different cells the nematode has: 959. They know how many are nerve cells — 302 — and they know which of those neurons are linked to which. It is the only animal for which we have a complete neuronal wiring diagram, for which we can say we know what every neuron looks like, how it branches, how its dendrites and axons interthread with other nerve cells. Scientists also know that during larval development, exactly 131 cells arise through cell division only to die within thirty minutes of their debut. The cells seem genetically fated for immediate destruction. That make 'n' break approach may sound wasteful, but scientists speculate that the doomed cells permit surviving cells to fulfill their destiny. Whatever the reason for their cameo appearance, the cells are proving useful in the identification of the genes in charge of cell death. Such results may eventually improve our understanding of the massive cell degeneration found in Alzheimer's, Parkinson's, and other diseases of aging.

Much of our understanding of nematode biology rests on the animal's most outstanding trait: its transparent flesh. Peering into the nematode's see-through body, geneticists can watch the cells divide and divide again at every stage of development, from fertilization to maturation. They know that all of the worms have the same number of cells; they know which cells give rise to which, and they can draw up lineage maps indicating the genealogy of each of the 959 cells and its ultimate fate as part of the head, nervous system, tail, vulva, or elsewhere.

People who study worms often lapse into a lover's rhapsody about the objects of their research, even after having spent years staring at them through microscopes. Mouse biologists don't do that, nor do lords of the fruit fly. Because human geneticists spend their time studying human diseases, they rarely wax poetic about what a piece of work is man, and besides, most of them are too busy trying to get patents on their discoveries. But those in the C. elegans field talk dreamily about the grace of worms, the elegance of worms, the way you can almost cuddle up to the squirmy little parentheses, and the thrill that never ceases when you watch a growing worm and see its cells divide. One biologist recalled a course he taught on worm genetics. He and his students were sitting around at midnight on a Saturday, watching a video that had been taken of nematode embryogenesis. A dozen people huddled around the screen, cheering, whooping, and gasping every time a cell divided in two. Somebody came along and said, hey, there's a party down the block, but the students waved the interrupter away. They didn't want to miss a single throb or shudder of a single growing cell.

The beauty of the beast is only part of what drew Dr. Sydney Brenner, of the Medical Research Council in Cambridge, England, to begin using nematodes as an experimental animal in the 1960s. Seeking to study problems of neurobiology and development, he wanted an organism that could become the multicellular equivalent of the E. coli bacterium, which was then the experimental workhorse of molecular biology. Scouring through zoology textbooks, he settled on C. elegans, not just because it was transparent and thus easily observed, but because it combined simplicity and complexity.

THE VERY PULSE OF THE MACHINE

Small as it is and though its neurons are few, the worm has a sizable repertory of behaviors. It will move in a sensual sine wave pattern toward odors that attract it, like those in waste deposits signaling the presence of a potential bacterial meal. It will shun places with excessive pockets of salt, which can dry it up. The nematode can sense temperature; it avoids being touched; and it has a biological clock that reminds it to defecate every fifty seconds. Male worms devote almost as much time to seeking a mate as they do to seeking food.

Of equal importance, the worm has a sexual system that makes for exceptional malleability. There are two types of worms, males and hermaphrodites. Males can inseminate hermaphrodites, which carry eggs, but what makes the hermaphrodites valuable to biology is that they also come equipped with sperm. When no males are available, the hermaphrodites can fertilize themselves. Because of that feature, Dr. Brenner realized he could manipulate the genetic information of a hermaphrodite without worrying about upsetting its mating skills; and the ability to tinker with genes and observe the effects of mutations on an organism's development and behavior is a key to success in genetics. Using worm hermaphrodites, geneticists can study even the most bizarre mutations passed along from one generation to the next. They can interfere with the musculature and the nervous system to the point where a worm can't move but can still reproduce, thus offering up an endless crop of baby mutants for analysis.

The people Dr. Brenner chose to work with quickly became converts to the new church of the roundworm. The most devout was Dr. John Sulston, who for ten years in the 1970s and early 1980s put his life on hold to chart the famed *C. elegans* lineage map. Hunched over a microscope, he observed and recorded each and every cell division in the complete roundworm life cycle. Deep tracks show in the floor of the laboratory where he rolled his chair between the dissecting microscope, where he mounted and prepared the worms for viewing, and the observation microscope, where he watched cells unfold. Younger nematode researchers visit the lab as though it were a shrine, and point to the floor marks as one might to the signs of a weeping Virgin Mary.

Researchers have taken the detailed information about cell lineage one step further by studying the genes that determine a cell's fate. That work has led them to identify a wealth of worm genes that look like mammalian genes. The fact that nature has conserved these genes over billions of years of evolution strongly suggests that they are indispensable to cell life and health. Among the genes considered most critical to controlling cell growth are the oncogenes. In their normal state, these oversee cell division or maturation, but when they are mutated by carcinogens, oncogenes can cause cancer, at least in higher animals like us. The nematode work helps illuminate the mechanism of the oncogene in both its normal and disturbed states. For example, the worm has its own renditions of two genes thought to be involved in many human tumors, one called "ras" and the other, the epidermal growth factor receptor. It turns out that in the hermaphroditic nematode, the genes perform an indispensable function, working in tandem to foster the growth of the genitals. First, the receptor gene switches on inside three cells on the creature's budding torso. It then sends a signal to the ras gene buried inside each of the three cells, setting off a tumultuous cascade of activity. Fifty other genes leap to life, and these tell the three budding cells to divide into the vulval structure, an opening between the animal's exterior and its ovary. At least, that's what happens in a healthy roundworm. If, however, the worm's ras gene is experimentally mutated to mimic the human cancer gene, the hermaphrodite will grow a sort of genital equivalent of cancer, ending up with multiple vulvas. This fantastic outcome not only offers researchers a handle on how the ras gene operates; it also has provided fodder for social events. "I notice that whenever I'm in a restaurant and mention these multivulval structures, the conversations around me quiet down," a scientist told me. "Everyone is listening, wondering what we could possibly be talking about."

Other comparative studies have given us wormly equivalents of the human gene for insulin, the human gene that makes an essential component of muscle cells, and the human gene for a protein that helps tug apart a dividing cell. The first gene to be found that gives a primitive, shapeless cell its initial nudge toward the form and function of a nerve cell was found in the nematode. If we have any hope of understanding the nervous system, we must root out the signals that make it nervy to begin with.

For all the extravagant praise that nematologists heap on their model organism, they predict that the best discoveries are still in the making. The cellular and neural anatomy of the worm has been mastered. Its physical map is almost complete, and its genetic sequence will follow shortly. *Caenorhabditis elegans* may be born in mud, but its future is assured as biology's version of a studio star.

10

THE WRAPPING

OF DNA

IF STRETCHED to its full length, a single molecule of human DNA would extend more than three feet, the height of the average nursery school child. But when squeezed and coiled and crammed into its rightful place in the bosom of the cell, the molecule of life measures about a hundred-thousandth of an inch across.

The extraordinary feat of packing long, viscous ribbons of genetic material into a spot too tiny to qualify even as a Manhattan closet is performed largely by histones, a family of five proteins that hug the DNA and condense it to size. Yet as considerable as this compaction job may be, histones have long been dismissed as dull structural elements, little more than the biochemical equivalent of nuts, bolts, and bungee cords that align the all-important genetic molecule in its proper dimensions.

Now evidence is mounting that the proteins do far more than compress DNA. Histones have proved to be intimately involved with one of the central tasks of staying alive: that of switching genes on and off along the DNA molecule in a deftly timed and delicately patterned fashion. It is through the control of gene activity that each cell performs its characteristic task — secreting bile salts if it's a gall bladder cell or metabolic hormones if it's a thyroid gland cell. To control gene activity, histones act as jealous competitors against other proteins for the allegiance of DNA. In some cases, when proteins known as transcription factors approach the genetic molecule determined to grasp it and set off a round of gene activity, the proteins must first outbid the histones that are already there, arms folded, fierce as bodyguards. Only under the most exacting of biochemical conditions will the histones cede enough space to let the supplicant transcription factors massage the DNA and thus get their message across.

In one cell type, the histones may loosen up a stretch of the chromosomes, thrusting it out toward the world in preparation for business; in another cell, the same region may be shuttered up, tucked away, as good as dead. Histones seem to be particularly critical for keeping genes turned off. Although most people think of inactivity as a null state requiring nothing more than a lack of input, it turns out that some genes, if not firmly repressed by histones, would be kept working at a low and potentially lethal level in the cell. Just being switched off is an active process, and histones work as industriously to muzzle the genes as the Dutch boy worked to keep the dam from breaking.

By understanding how the small clan of histone proteins helps set the tone and amplitude for gene expression in the trillions of cells of the body, scientists hope to gain a clue to the great puzzle of how cells know what they are, when they should divide, and when they must expire.

Knowledge about histone behavior and the architecture of DNA may eventually reveal the mechanisms behind some diseases. For example, common blood disorders known as thalassemias sometimes stem from a distorted twisting in the structure of the DNA molecule, possibly as a result of histone irregularities. And because cancer is a disease of abnormal cell division and aberrant gene activity, the disturbance of the histone proteins that protect DNA is thought to be a significant step in malignant transformation.

But, for reasons more immediate than the search for any utilitarian value, scientists are interested in histones because the proteins have given them yet another rich subject to quarrel over. Not everybody is prepared to accept that they are worth anything beyond a dismissive grunt, whatever the data may say. Much of the resistance to histones is historical. The revolution in our understanding of how genes work began with the study of bacteria like *E. coli*, whose DNA is packaged very differently from that of the cells of higher organisms. Bacterial DNA is loose in the gelatinous sea of cell, not protectively sequestered in a nucleus at the cell center, nor is it enmeshed in histones. If bacteria can make it without histones, why bother fussing over histones? Even when scientists chose to study the DNA of higher creatures, they often began their experiments by chemically defrocking the genetic molecule of its surrounding histones and tossing the naked strand into a test tube. They then studied the behavior of the special proteins that copy and activate the DNA.

That pristine test-tube method has yielded knowledge about the signals that spark the copying of different genes, but lately a number of scientists have derided the approach as reductionist, misleading, and in some cases wrong. They argue that DNA in its histone packaging behaves very differently from the stripped-down DNA that is studied *in vitro*, in glass tubes, and they insist that animal DNA must be considered in its native state, in the nucleus, with all its confounding embroidery.

By picking apart the genetic material as it is structured in chromosomes, researchers have learned that at its core it is like a chunky costume-jewelry necklace, built of equal parts of histone and DNA. Forming the beads are four pairs of histone proteins, merged into a little particle known as an octomer, which bears a striking resemblance to Mickey Mouse. A strand of 146 bases, or subunits, of DNA is wrapped twice around the octomer, producing a nucleosome, the fundamental unit of the chromosome.

What links the individual nucleosome beads is a series of short braids comprising skeins of the fifth variety of histone protein, interwoven with fifty more bases of DNA. Those alternating lengths of beads and chains are wrapped and coiled and tightened and squashed around and around one another, with many ancillary nonhistone proteins becoming attached to the necklace. It is this arrangement that brings about the remarkable packaging of a meter-long molecule into an invisible speck, and it is this arrangement that unites all creatures more complex than a germ. Whether you look at the cell nucleus of a person, a yeast cell, a bumblebee, a turkey, or a corn plant, you'll see the same histone proteins, squeezing the chromosomes breathless.

Yet even with such detailed revelations about histone structure and evolution, proponents long had difficulty making their case that the proteins were anything other than inert molecules, with all the charm, as one put it, of a chair leg. Only with the introduction of several technical advances have researchers begun to see when, how, and under what circumstances histones control the genes so gracefully wrapped around and between them. One breakthrough is the adroit manipulation of yeast cells in the laboratory so that histone production can be modulated at will. In another bit of legerdemain, scientists have managed to re-create in a test tube all the interacting elements of the nucleosome beads and linker chains, in essence simulating the living nucleus but under well-defined laboratory conditions.

Through such advances, it has become clear that different regions of histones can tickle the DNA in a manner far more exacting than anybody had thought possible. One study of yeast cells showed that if a single small element at the beginning of a histone protein is destroyed, the cells survive, but barely: with histone production thrown out of whack, several genes needed to metabolize the fungus's sugary meals cannot be switched on. If, however, the opposite end of the same histone protein is disrupted, the cells can no longer turn a few essential genes *off*. The complementary studies offered the first clear evidence of histones' subtle and versatile talents for orchestrating gene behavior. One end of the protein is a torch for lighting genes up, the other a candle snuffer for blacking out the flame.

·   ·   ·

In several test-tube simulation studies, researchers have shown that histones aggressively vie with other proteins, the transcription factors, for the privilege of attaching to DNA. The results of those contests vary considerably, depending on the genetic sequence under study.

In some instances, the histones seem to be bound quite casually to the DNA molecule, and will relinquish their position readily when the proper transcription factors are introduced. These factors are proteins specifically designed to catalyze the gene, which otherwise would slothfully do nothing at all. In other instances, the histones seem more solidly entrenched in their position along the double helix. The competing transcription factors seem designed strictly to shove the histones aside, at which point the gene attached to it, by its very nature rambunctious, becomes active. Sometimes, both the histone protein and its competitor are hooked on to the DNA simultaneously, apparently as a way of activating the gene — but only modestly or in spurts.

It's not surprising that evolution would use histones to perform a broad variety of tasks, although some may not have been in the initial job description. Histones probably began their service a billion or so years ago as packaging cords, to ensure the orderly array of the chromosomes within the nucleus; but they have since taken on far more illustrious responsibilities as conductors of gene activity.

The higher the rank, the heavier the fall. There's good evidence that disruption of the nucleosome structure in our cells can have devastating consequences. One dramatic example is a group of anemia disorders common among Mediterranean people. Most types of thalassemia result from mutations in the genes that allow blood cells to make hemoglobin. But there's one version of the disease that results from a flaw, not in the hemoglobin gene itself, but in the ability of the chromosome where the globin gene sits to unfold into its proper position, after which it can be cajoled to synthesize hemoglobin. Instead of being floppy and turned outward, as it should be, the chromosomal spot looks like an angry snarl of hair, too densely packed to be reached by activating factors from the cell. At the root of the problem is an aberration in histone architecture, a distortion in the proteins charged with protecting and disciplining the genes around them.

As scientists delve deeper into the nuances of chromosome structure, they confess they run the risk of again oversimplifying the story of the cell by reducing all gene problems to histones. In the supple, restless dynamo that is the living cell, no single family of proteins can do everything needed to keep entropy at bay.

11

CHAPERONING PROTEINS

WHEN A NEW PROTEIN slides off the tiny molecular as-
sembly line within the cell, it is nothing more than a droopy
string of amino acids, not yet fit for its designated profession.
Only after being spun and pleated and braided into its proper
three-dimensional conformation will a protein burst to life, seiz-
ing up oxygen if it is hemoglobin, shearing apart sugars if it is
an enzyme, or lashing cells together if it is the stout twine of
collagen.

Until recently science had scant idea how a simple chemical
strand manages to fold into a working protein, with all its knobs,
clefts, sheets, and curves arrayed in vivid harmony, able to min-
gle with other molecules around it. The problem is no mere
academic exercise, but a question of central importance to biology. Proteins perform most of the tens of thousands of tasks
needed to keep the body alive, and only a perfectly folded protein
is up to the demands made upon it. Protein folding was thought
to happen spontaneously, a newborn protein, or polypeptide,
springing into its correct three-dimensional shape on its own,
driven solely by the repellent or attractive electrical and chemical
forces of its individual amino acids. As scientists tried to calculate what those enormously complicated interactions might be,
they made little headway in cracking the folding conundrum.

So it is a great surprise to discover that nature doesn't let proteins fold up by themselves but has created a whole family of proteins whose sole purpose is to help other proteins crinkle and furrow. The detection of the handmaiden proteins, called chaperones, means that the traditional theory of spontaneous folding is mistaken: the forces inherent in a polypeptide's sequence of amino acids aren't enough to sculpture and knead a protein into its correct, muscular form.

Instead, protein folding, like so much of what happens in the body, turns out to be done by committee. As the amino acid chain rolls forth from its birthing chamber in the cell, folding must begin immediately, and to that end, successive bevies of chaperone midwives rush over and gently embrace the flat polypeptide at hundreds of key spots, shielding it against the hostile environment of the cell. The chaperones allow those amino acids that are destined for the interior of the active protein to curl in on themselves, while they encourage those regions meant for the exterior to turn and face the outside world. They help twist some stretches into corkscrews and pummel others into flat sheets. The chaperones also protect the fragile chain from becoming ensnarled with other infant peptides floating in the cell, as it would if left unattended.

Nor do the jobs of chaperones end once the initial folding is through. Should the cell suffer a shock from extreme heat, oxygen cutoff, or any sort of trauma that threatens the structural integrity of the thousands of proteins within, the chaperones will toil mightily to prevent protein disintegration, latching on to the wilting molecules and helping to bend them back into shape. So indispensable are the folding molecules to growth and survival that cells experimentally deprived of their chaperones rapidly die.

The information that researchers are gathering about chaperones could yield knowledge about many of our worst afflictions. After a heart attack, for example, large patches of cardiac muscle atrophy and die as a result of temporary oxygen deprivation. If the chaperones in heart tissue could be manipulated right after a heart attack, the restorative molecules could shore up the collapsing proteins in the heart cells and perhaps prevent tissue death. Certain genetic disorders, like muscle-wasting diseases, may result from mutations that slightly weaken the cell's chaperones, leaving many proteins in disarray. In theory, a better understanding of chaperones will improve drug design. Many drugs currently in use and under investigation are based on natural proteins. If pharmaceutical companies could learn why a particular amino acid prompts a protein to twirl up rather than down, and why the protein works better in one shape than in another, they could mix and match components to improve on nature's offerings.

Before we get swept away with easy enthusiasm, though, we should bear in mind that chaperones are only a part of the protein-folding puzzle; much remains to be learned about the wildly complex dynamics of protein structure. The information needed to determine the final working profile of the protein is inscribed in the sequence of its amino acids, and scientists still do not understand the electrochemical pushes and pulls of those building blocks. Chaperones do not dictate how a protein folds but only help the protein realize its ambitions and steer it away from binding with bad company. In other words, they're the foremen of the floor: they make sure the job gets done right, but they're not the ones who decide what the final product should look like. Those specs, written in the amino acid code of the protein, have yet to be deciphered.

Nevertheless, the discovery of chaperones is a bright spot in an otherwise discouraging discipline. By tracing the interactions between a folding protein and its industrious assistants, scientists just may be able to identify all the intermediate steps between a flaccid polypeptide and a strapping folded protein.

The reason that scientists so long neglected to recognize chaperones is that they carried out their studies of protein folding *in vitro,* throwing together isolated protein subunits and a few other ingredients to see what emerged. Biochemists found that nearly any polypeptide they tossed into the test tube would, if cultivated under exacting conditions, fold into its active shape as the different amino acids on the chain looped in one direction or another, depending on their inherent molecular properties. This discovery led researchers to assume that protein folding happened spontaneously in the cell as well. In an effort to understand folding dynamics, they used intricate mathematics, the principles of physics, computer graphics, and difficult crystallography techniques, but with only modest progress.

In the mid-1980s, scientists who worked not with isolated molecules but with living cells detected several proteins that flared into action when the cells were subjected to abnormally high temperatures. Those responders, named heat-shock proteins, were determined to play a crucial role in helping the cell weather the heat by stabilizing all the rest of the proteins, which otherwise would unravel.

Biologists, after identifying many kindred of the original salvation proteins, divided them into at least two superfamilies of proteins and observed them in creatures across the evolutionary landscape, from bacteria to humans. A big advance occurred when those same heat-shock proteins were spied in normal cells that had not been stuck into a laboratory oven, suggesting that stress proteins participated in the daily life of the cell and did not serve only as an emergency crew.

Geneticists then discovered that yeast cells harboring mutations in their heat-shock genes were in terrible shape, a mess from their nuclear heart to their rubbery membranes. The proteins in these cells hadn't folded properly, a defect that led to wholesale havoc. The biologists realized they had an unexpected bonanza: a class of proteins to cast light on the blackness of the folding problem. That is when the proteins were rechristened chaperones to reflect their more general duties, although they are also called stress proteins, GroEl, or any number of unevocative names.

Most of the chaperone experiments to date have been performed in yeast or bacterial cells, which are easily manipulated, or in isolated cellular structures, like the mitochondria, where the body's energy is produced and where many proteins must be created and folded to help stoke the power plant. Researchers now understand that, from the perspective of a newborn polypeptide, conditions in a living cell differ dramatically from those in a test tube, and that chaperones serve as indispensable nursemaids. One type of chaperone after another steps in to assist as folding begins, a process that takes an average of three or four minutes. "It's like Snow White and her seven dwarfs," Dr. Mary-Jane Gething told me. "One dwarf has the hammer, another the chisel, a third the shovel, and so forth."

As a stretch of amino acids begins rolling off one of the cell's ribosomes, the pear-shaped factories where proteins are created, a small chaperone called hsp70 drifts over and grips certain tender areas of the polypeptide. The chaperone recognizes stretches of amino acids that are hydrophobic, or water-hating. Such hydrophobic patches are destined to end up tucked inside the protein once folding is finished. But until they can curl under, they are vulnerable to misguided merging with other polypeptides, and thus must be safeguarded by the chaperones. The concentration of proteins in the cell is as thick as honey, and young proteins must be sequestered from the ambient ooze. During the early stages of folding, the polypeptide may form characteristic corkscrew shapes, or linked loops that resemble a Christmas bow, or slender fingerlike projections. When the preliminary folding is complete, members of the first round of chaperones relax their grip and drift off.

As folding proceeds, and the bows and corkscrews really begin to twist in on themselves, a second group of chaperones, called hsp60, takes over. This molecule looks like two doughnuts stacked on top of each other, offering a snug tunnel into which the partly folded protein is pulled for further twisting without interference from stray peptide strands outside. Eventually, the protein torques into something round, dense, and energetic, perhaps sporting a pincerlike cap to snare hormones or a deep groove to capture a foreign microbe. Upon the completion of folding, the chaperones let the protein loose to try its luck in the thick of life and move on to the next newborn in need of care, approaching and abandoning the whorls and loops and jags of an ever-tightening peptide chain. Hundreds if not thousands of times each hour, they are alchemists, spinning dull chemical straw into a splash of protein gold.

12

A CLUE

TO LONGEVITY

HUMAN CHROMOSOMES, shaped like cinch-waisted sausages and sequestered in nearly every cell of the body, are famed as the place where human genes reside. But a few small architectural details of the chromosomes merit at least as much celebrity as the 100,000 genes they hold. At the very tips of the chromosomes are extraordinary structures built of six DNA letters repeated over and over, thousands of times, like monotonous molecular chants. The monotony belies a song of songs, for these structures, called telomeres, declare their importance at every stage of the cell's life. They shield the chromosomes against harm, organize them into their proper position within the nucleus, and, most provocatively, serve as a sort of timekeeping mechanism that tells the cell how old it is. Each time one of our cells divides, the telomeres are shaved down by a fixed and ever so slight amount. Thus, the length of the chromosome tip can offer the cell a measure of how many times it has divided — and how many divisions remain before the cell's life span is ended. Although individual cells die and are replaced throughout life, there is a strong correlation between telomere length and the decline of the entire human being. On average, the telomeres of a seventy-year-old are much shorter than those of a child, and it's quite possible that once the telomeres of most cells in our body fall below a certain length, we are very near the end. Small wonder, then, that the work on telomeres has broad implications for the study of aging; some scientists have proposed, for example, that patching up the chromosome's telomeres could help shore up elderly cells, particularly in parts of the body where wear and tear are most relentless, as they are along the coronary arteries.

Apart from filling their role as the cell's hourglass, telomeres undergo a dramatic transformation during the genesis and progression of cancer. When a cell becomes malignant and begins dividing out of bounds, its telomeres are shortened, whittled down with each illicit cellular split. Telomere length thus offers a means of judging the stage of a cancer cell: the more diminutive the tip compared with the patient's healthy cells, the more advanced the malignancy. At a very advanced and highly aggressive stage of tumor development, however, an ugly change can occur — the telomeres may stop shrinking and begin to grow again. Studying cancer cells grown in laboratory dishes, scientists have determined that the vast majority eventually die but that a fraction become immortal, essentially able to divide forever. These cells, it turns out, contain an enzyme called telomerase, a builder of telomere tips. Telomerase normally stays quiet in adult cells, but in the immortal cells it has somehow managed to swing back into action and start reconstructing the telomeres, a talent that can prove lethal for a cancer patient. If telomere length is the timepiece designed to tell a cell to go quietly, and if that length fails to dwindle from one doubling to the next, there may be no internal signal to prevent the cell's ambitions from overwhelming the entire body.

The recent knowledge about telomerase suggests a way to block late-stage cancer. If a drug could interfere with the activity of the telomerase enzyme, the immortal cancer cells could be destroyed without damage to most normal cells. Searching through a variety of healthy tissue types, biologists have detected hints of telomerase activity only in sperm cells, where the enzyme keeps chromosomes at the most youthfully elongated length possible.

Thus, an attack targeted at telomerase may have no more severe effect than the lowering of sperm production. Designing such a blocker may not be difficult. Telomerase chemically resembles reverse transcriptase, the enzyme at work in the virus that causes AIDS. Several drugs already exist that interfere with reverse transcriptase, among them AZT and ddI, and the possibility that these compounds may block telomerase as well is now under exploration.

Whatever their practical application, telomeres speak in their loquacious, redundant fashion to the more fundamental issue of chromosome design: how the genes are packed into the cell and why they're positioned where they are. Science knows almost nothing about the large-scale organization of the genes, and telomeres offer a few tantalizing hints. For example, the telomeres somehow prompt the genes seated next to them to recombine at an unusually high rate, mixing and matching and switching pieces during the creation of sperm and eggs. That high recombination rate would be desirable for genes demanding great variability — say, the genes of the immune system — and much less welcome for genes that require great stability, like those which take care of humdrum housekeeping tasks inside the cell. It is the genes near the telomeres that are thought to be the changelings, our most mutable genes, perhaps evolving at a slightly accelerated rate relative to the rest of our genome. In a sense, the neighborhood of the telomeres may tell us of our evolutionary future, though reading those particular tea leaves remains beyond our current skills.

The era of the telomere began in the early 1970s, when one of the greats of modern genetics, Dr. Barbara McClintock, of Cold Spring Harbor Laboratory, observed during her studies of corn that broken chromosomes were extremely unstable, likely to snap into smaller pieces and eventually cave in on themselves. She proposed that normal chromosomes must possess protective structures that prevent their degradation. Those structures turned out to be the telomeres. Others took up the detailed investigation of these chromosomal guardrails, turning to single-cell organisms similar to the paramecia that swim through pond water. Such creatures have a distinct experimental advantage. Whereas each of our cells has only forty-six chromosomes, with a telomere tip at either end, the pond dwellers carry tens of thousands of tiny chromosomes, every one tipped at top and bottom with a telomere that can be isolated and studied.

By examining the telomeres in fine detail, investigators determined that they are composed of a sequence of only six DNA bases, or chemical subunits, repeated thousands of times at both ends of every single chromosome. In our cells, the sequence is TTAGGG, and it is repeated anywhere from a thousand to three thousand times, depending on the cell type and its age. The sequence does not mean anything on its own, but, among other tasks, it serves to stabilize the chromosomes. Like bookends, the telomeres hold everything in place.

For the single-celled pond creatures, which divide indefinitely, the telomeres themselves must also be held in place; the paramecia can't afford to have their telomeres dwindle down every time they give a wriggle and split across the middle, because they'd eventually reproduce themselves into extinction. The cells of pond scum happen to contain abundant amounts of the repair enzyme telomerase, which itself looks like nothing scientists have ever seen before. Tucked within the enzyme complex is a little chemical rubber stamp, an RNA template made of the six letters of the telomere. The enzyme uses that template to stamp a new sextuplet of bases to the tip of the telomere and replace any bits of the bookend that were lost during cell division.

Researchers have since observed that, while the telomerase rubber-stamp enzyme works in paramecia and other simple species full time, it is not active in most of our tissue or in that of other mammals. Instead, our telomeres shrink with age. Normal human cells divide between seventy and a hundred times, and with each split they lose about fifty letters of their telomeres from both chromosomal ends. Before the telomeres vanish altogether, the chromosomes sandwiched between them become warped and sticky and attach themselves to other chromosomes in such twisted configurations that the cell has no choice but to die. Granted, it's possible that telomere degradation does not cause cell aging and death but merely accompanies it, rather like gray hair or presbyopia. Still, there is some highly suggestive evidence that telomeres do keep track of the passage of time and set off alarms at the final moment. Studies of cells proliferating in laboratory dishes have shown that the shrinking telomeres maintain the chromosomes in a remarkably stable state until right before the cell's allotted time is over, rather than letting the whole mess gradually disintegrate, as might happen if the telomeres were incidental to the decay. Instead, extremely short telomeres seem to signal other molecules in the cell to stop working, stop trying, stop dividing, and prepare for dying.

Yet the link between chromosome tips and aging remains tenuous, and no miraculous telomeric anti-aging potion is likely to emerge from the research any time soon. Nor should we want to fool around with this sort of drug. After all, we've seen what happens when the cells of an aging adult suddenly assume the telomeres of youth: they keep on growing, with the pedal to the floor.

13

WHAT HAPPENS
WHEN DNA IS BENT

By TRADITIONAL RECKONING, DNA is the benign dictator of the cell, the omniscient molecule that issues commands to create enzymes, metabolize food, or die a willing death. Recent advances, though, suggest that DNA is more like your average politician, surrounded by a flock of protein handlers and advisers that must vigorously massage it, twist it, and, on occasion, reinvent it before the blueprint of the body can make any sense at all. Scientists have found an extraordinary class of proteins that serve almost exclusively as molecular musclemen, able to grab the cell's genetic material and, in a fraction of a second, kink the strands into hairpin curves. Then, just as quickly as they flexed the molecule of life, the proteins jump off and allow the DNA regions to snap back to straightness.

The act of bending DNA plays a critical role in controlling genes. In some cases it brings together far-flung bits of genetic information into a working command for the cell; at other times it goads teams of proteins to join forces at one well-bent spot and galvanize biochemical activity. By momentarily scrunching up parts of the chromosomes, the benders seem to influence the verve of the immune system, the sex of an infant, the scrimmages between a virus and its host. And the proteins add new proof to scientists' growing conviction that the complex architecture of

DNA is at least as crucial to the behavior of genes as is the sequence of chemical letters of which the genes are composed.

As the most persuasive evidence of the importance of bending, a protein celebrated as the key to masculinity performs its magic by acting as a potent DNA bender. The protein, testis-determining factor, was first identified in 1990 as the long-sought maleness signal, a molecule that somehow sets off a cascade of biochemical events and helps transform a fetus of indeterminate gender into a baby boy. Since then, we've learned that the factor accomplishes its task by manipulating DNA at many spots up and down the chromosomes, bending by almost 90 degrees whatever straight region it concentrates on. Through a sequence of bends, molecules that otherwise are dispersed uselessly across the double helix can be brought into contact with one another. And, once joined into squadrons by the go-between bender, the molecules become active switches — the transcription factors and enhancers — able to mobilize a battery of genes and turn them into new proteins and enzymes. These proteins can shape a fetus's genital buds into little testicles, the pivotal step in the making of a male. So, if boys are accused of mangling everything they get their hands on, they can blame their genes, which after all had to be bent beyond recognition before they could be boys.

Another protein, lymphoid enhancer factor, influences the production of the T cells of the immune system. That protein is an even more emphatic DNA bender; it crimps angles of 130 degrees into the chromosomes and, like the testis factor, introduces distant transcription molecules to one another, igniting gene activity. In this case, the stimulated genes help to create T cell receptors, proteins, extruding from the surface of immune cells, that can recognize nearly every foreign element storming the body.

The bending of DNA is also involved in more nefarious events, like the virus's invasion of a host's chromosomes. Researchers studying the complex interplay between a bacterial cell and its parasitic virus, the phage, find that the virus gets into the host chromosome by exploiting the bacterium's own DNA-bending proteins, generating crimps in the DNA, and then sneakily cutting and pasting its genetic information into the curled-over sequence.

Many human viruses, notably the one that causes AIDS, are thought to adopt a similar ruse, bending the DNA and then integrating permanently into the chromosomes. Bending DNA can also kill cells. One widely used chemotherapeutic drug, cis-platin, destroys the rapidly dividing cells of tumors by bowing the DNA into an abnormal configuration. That bending in turn attracts other proteins to the pleated areas, a molecular caucusing that ends up blocking DNA replication and killing the cell. By understanding the kinetics of normal and pathological DNA bending, biologists may be in a better position to devise strategies to foil integrating viruses or effectively to jam the gears of a tumor cell.

Above all, the revelations about DNA bending emphasize the importance of the dynamic architecture of DNA to its performance. The emerging work on benders and other proteins that tweak, sheathe, wiggle, and whet the double helix demonstrates that DNA is in continuous motion and forever communicating with the throngs of molecules around it. That ceaseless activity and flexibility means that the chemical instructions encoded within the individual genes can be read in a surpassing variety of ways. "The superstructure of DNA is dynamic," one scientist says. "It changes with time."

As with so many discoveries in the field of gene control, much of our current understanding of DNA bending comes from bacterial studies. Bacterial DNA is shorter and simpler than mammalian DNA, and all its parts and switches are designed for optimal performance, to permit the microbes to replicate as swiftly as possible. Scientists studying the amenable DNA soon realized that although it has, when considered in its entirety, many bends and curves — hence the image of a winding helix — short regions of it are quite stiff. To help me see the distinction, a scientist suggested that I think of a long stick, like the bow of a bow and arrow, which is highly flexible. If I were to try to bend any short bit of that stick, though, I'd find it behaving like a rigid rod. Similarly, DNA in its superstructure is extremely plastic, yet any little part of it stays taut.

Scientists, however, learned early on in their investigations that the tautness of even short stretches sometimes relaxes. For example, when certain sequences of DNA occur together, like a repeated string of the chemical base adenine, that region of the chromosome ends up curving over slightly. Delving further, they noted that those curving regions happen to be sites of interesting activity. Strings of adenine represent parts of the chromosome where gene transcription begins: where DNA is written into another chemical form, RNA, the first step toward the creation of proteins. Somehow, the inherent chromosomal floppiness that comes with a lineup of adenines helps arouse genetic activity.

Other work divulged that when a virus sought to infiltrate the genes of a bacterium, it managed to bend the region of the host chromosome and then stitch its own genes into the kink. The finding led to the isolation of a bacterial protein called integration host factor, or IHF, which can bend DNA by 140 degrees, or almost back over on itself. That bacterium normally uses the bending factor to mix together, or recombine, its genetic material during ordinary cell division, so the crafty phage virus uses the IHF protein for its own ends. It fakes the bacterium into thinking its DNA is bending simply for the sake of healthy recombination, but then the virus slips its own genetic luggage into the wrinkled region.

Scientists subsequently identified a giant family of proteins related to the integration host factor, and have been systematically investigating the molecular tribe to see which ones can bend DNA and for what purposes. Some of the proteins head for specific sequences of DNA and then sharply deform the region. Others are more profligate benders, contorting almost any patch of chromosome they alight on.

Two broad themes have emerged to explain why bending has a dramatic effect on DNA. For one thing, the act of turning on any gene and translating it into a newborn protein is complex; nothing works by a mere on-off switch. Preceding a gene are strings of genetic sequences that allow the cell's protein-manufacturing equipment to attach to the DNA and work at a given speed and force. Those instructions often are spread out over dozens to hundreds of bases, and scientists now believe that the DNA is bent and furrowed to put the otherwise nonsensical words into proper order.

Sometimes DNA must be bent, not to bring genetic sequences into proximity, but to permit contact between proteins that cling to the double helix like barnacles to a whale's back. Often these proteins need to meet their partners before they can perform their duties, which include tending the DNA molecule, repairing any errors that inevitably arise along its length, replicating it into a whole new strand before cell division, or doing something entirely unrelated to DNA upkeep. In such cases, bending proteins will strong-arm the DNA to let the proteins touch, and the illustrious double helix becomes nothing more than a kind of Rube Goldberg transportation device to get two proteins together. It's like a floppy disk for a computer, one biologist said. Normally, the floppy disk, the DNA, feeds instructions to the computer — it's the software of the cell. But if you were to use a floppy disk to, say, prop up the leg of your desk, the software would for the moment function as a bit of hardware.

In bending the DNA to their bidding, the proteins do not exactly grab it in their fists and give a twist. Rather, like bowling balls tossed onto a trampoline, they cause a deformation in the linear structure of DNA.

While bending proteins obviously are important to the bacterial cells in which they were discovered, the protein family is likely to be even more critical to so-called eukaryotic cells, the components of higher creatures like us. That is because the DNA of higher creatures is wrapped in protective packages of proteins, notably the histone proteins, which must be elbowed aside quite firmly so that the genes within can burst to life. The bending proteins seem to compete with histones for the privilege of latching on to DNA, but unlike the histones, which pack it tightly away, the benders jostle it into action.

What's more, the intricacy of gene control in animal cells far exceeds that in bacteria. There are controls that control the controls in charge of the genes — transcription factors, enhancers, promoters, a teeming bestiary of switches and modulators that allow us to be all that we are without even giving it a thought. The perpetual flexing of DNA may be the easiest way for the cell to keep its restless fauna in line.

14

BLUEPRINT

FOR AN EMBRYO

As any ardent viewer of the "Star Trek" television series will attest, the majority of alien creatures portrayed, no matter how theoretically distant their origins, look reassuringly familiar. They have bodies separated into two basic regions, head and trunk. They have eyes, ears, mouths, and snouts, though of varying puttied-up shapes and dimensions. They have arms and legs in tidy symmetrical pairs.

It is as though the show's producers believe in a Platonic ideal of a body plan, a way of putting an organism together that has such firm transgalactic logic behind it that the scheme has evolved independently time and planet again.

But lest the creators of the series be accused of intellectual timidity or laziness, we should recognize that nature too believes in the all-purpose, reusable blueprint for building mobile bodies. Researchers probing the earliest events in the transformation of a single fertilized egg into a breathing, sentient, multicellular being have discovered a class of genes that could well be the molecular signature of beasthood, the key to the kingdom Animalia.

These genes, the Hox genes, are among the potentates of animal development, working in the first few days of embryonic growth to lay out the basic structure and orientation of a body — where the head will be, where the limbs, the digits, the chest, where the organs sheltered within. Of great importance, Hox genes speak in many tongues: scientists first discovered them in fruit flies, but have since detected signs of Hox activity in the early embryos of mice, worms, fish, chickens, grasshoppers, humans, frogs, cows — indeed, in every creature studied to date.

The precise number of Hox genes within an animal's cells differs significantly between vertebrates like humans, who carry thirty-eight such genes, and invertebrates like fruit flies, which make do with just eight; but the fact that species separated by six hundred million years of evolution rely on the same class of genes to orchestrate embryonic growth suggests that the molecules work too splendidly to demand much tinkering or reinvention.

So primary are Hox genes to the design of the animal body that they may be considered the ultimate test of animalness, a potentially more precise definition than such traditional measures as independent movement and response to stimulation. The genes are not found in the cells of plants, fungi, or slime molds, and therefore scientists have proposed that the DNA of any species whose phylogenetic status is in doubt — for example, sponges and some groups of protozoa — should be examined for evidence of Hox genes.

These genes are not the sole players in the extravaganza of animal development; other molecules like steroid hormones, a derivative of vitamin A called retinoic acid, and a host of growth factors also participate in embryogenesis. But biologists have made great progress in cracking the puzzle of the Hox genes. They have created transgenic mice outfitted with telltale mutations in their Hox genes, and have artfully rearranged the influence of the genes on the budding wing of a chick embryo. Through such experiments, they have learned that Hox genes operate as master switches, producing the transcription factors that clamp on to the chromosomes and set off wave upon wave of subordinate genes, thereby amplifying a modest initiating signal into great crests of biochemical activity.

Hox genes essentially assign addresses to the cells of the early embryo, telling one cell it is a constituent of the front of the body and not of the rear, informing another that it is part of the limb destined to become a finger. They are vital for setting the body pattern from the hindbrain on down; other developmental genes are more critical for constructing the forebrain and related regions of the central nervous system.

The Hox genes operate with great speed and efficiency, performing their remarkable patterning task in three days, beginning sometime around the first week after conception. One reason they are able to help set up the whole design so speedily is that the embryo is constructed as a series of repetitious segments: a new section of the body grows out as a mimic of the portion preceding it, and variations are subsequently added to each section to lend the final body its complexity of parts. In other words, a baby grows rather like a venetian blind being lowered over a window, slat by slat. That segmentation of design can most clearly be seen in the redundant pattern of your spinal column and ribs, but other parts of the body also show a modular construction. From an evolutionary perspective, the scheme makes sense, for it allows nature to begin with one unit — the basic embryo segment — and then duplicate and modify, duplicate and modify, top to bottom, rather than starting from conceptual scratch for each organ. The Hox genes keep the modules coming as embryogenesis proceeds.

Another intriguing feature of the Hox genes is their structure, the startlingly precise manner in which they are organized on the chromosomes. The mystery of how the different Hox genes operate during embryonic growth is just beginning to be explained, but researchers think the genes collaborate with one another, sequentially flicking on as development proceeds, or in some cases working in teams to ensure the proper patterning of a certain part of the body. Such genetic collusion is hardly unusual, and many tasks of life require the simultaneous or consecutive contributions of multiple genes; for example, four different genes are needed to build a single protein cage of hemoglobin, which ferries oxygen in the body. But while the various hemo-globin participants are scattered hither and yon around the chromosomes and switch on as needed, no architectural organization required, the thirty-eight Hox genes in mammalian cells are grouped in four tight clusters on four different chromosomes. In each set, the genes are lined up, one after another, facing in exactly the same direction and with the same spacing between them, as though a designer had drawn the plan with a T-square and ruler.

Nowhere else among the 100,000 genes that make up human DNA have scientists found such an exact configuration, and some biologists believe that the singular arrangement is no accident. In one brazen proposal with enormous aesthetic appeal but scant evidence to back it up, French scientists suggest that the three-dimensional arrangement of the Hox genes on the chromosomes acts as a built-in clock, ticking off the moments as development unfolds by using the length and components of the double helix itself as the timing mechanism.

By this theory, one Hox gene turns on and produces a transcription factor that kindles many other genes to create the uppermost segment of the body; then the next Hox gene down is aroused, and it helps build the next lower section of the embryo, and so forth, all in lockstep timing. If this idea is taken to its logical conclusion, the Hox genes could be said to encapsulate the entire human form: the chromosomes contain a physical and temporal representation of the body axis, of the child itself — an idea with alluring artistic and philosophical undertones. It brings us back to the medieval notion of the homunculus, a tiny human being in every sperm cell that needed only the nourishment of the mother's womb to grow. Many depictions of the Annunciation, of the Archangel Gabriel telling the Virgin Mary she will bear the son of God, show a homunculus of Jesus headed toward Mary on a shaft of light. In our updated homunculus hypothesis, the little man or little woman is embedded in the four chromosomal groups of Hox genes hidden within the shadowy nucleus of the early embryonic cells.

Other developmental biologists are concerned less with the spatial arrangement of Hox genes on the DNA than with how the genes perform their magic and help form the pattern of the embryo. The molecules won their name some years ago when geneticists observed what happens when the genes are mutated in fruit flies. After bombarding their insects with blistering X-rays, researchers noticed that some of the flies' progeny emerged with startling anomalies, like double pairs of wings, two thoraxes, or long legs at the top of the head where their short antennae should have been. The researchers soon realized that the radiation exposure had essentially reprogrammed developmental genes, causing normal body pieces to grow where they did not belong; the legs springing up at the top of the fly's head were perfectly healthy legs, but a fly is not supposed to have legs coming out of its forehead. Mutations of this type were dubbed homeotic, meaning similar, because they caused similar body parts to sprout from more than one spot on the fly.

Scrutinizing these developmental genes in detail, biologists noticed that nested in each of them was an identical molecular sequence, which they designated the homeobox. The sequence is brief, a mere 183 base pairs, or building blocks, out of the thousands of DNA bases that make up individual genes. But the fact that the same motif was seen so often indicated the sequence must be a critical genetic player in embryonic growth. Thus, developmental genes that contain a homeobox were named Hox, after their core element.

Scientists have since learned that the homeobox is the so-called DNA-binding domain of the gene. When a Hox gene switches on and makes its protein inside the cell, the homeobox sees to it that the protein will have a flirty little corkscrew twist to it. That shape allows the Hox protein to curve around the double helix in a sinuous hug and activate other genes.

Still, studies of fruit flies, however essential to the story of the Hox genes, can go only so far in explaining the making of a mammal, and vertebrate biologists have ventured much further. Some are systematically knocking out the Hox genes in experimental mice to see how the creatures' development is affected by the absence of one or another of the thirty-eight players. They employ painstaking techniques to delete the Hox genes from embryonic mouse cells and then use those cells to breed lines of genetically deprived rodents. Depending on which Hox gene the experimental mouse must do without, it emerges from the womb with a singular type of birth defect. When Hox A-3 is deleted from embryonic mouse cells, for example, the pup is born with a broad suite of flaws: heart defects, face and skull deformities, and the lack of a thymus, the organ where immune cells mature. Although the deformities seem dispersed and unrelated, the work suggests that all the afflicted organs originated in the same segment of the early embryo. Without the Hox gene to stamp its initial design, that embryonic section developed in a grossly distorted manner, and all the organs it gave rise to suffered the consequences.

Many mysteries remain, however, far more than any that have been resolved to date. Among the major puzzles is determining which genes are activated by the Hox genes, and exactly what happens when the target genes respond. Do they turn on other genes? Do all the activated genes change chemical conditions in the embryonic cells? Do they increase the capacity of the flowering cells to respond to signals from their neighbors? Only by mapping out the entire cascade of events will scientists truly understand how the Hox genes work, and how one executive molecule like Hox A-3 can shape a sliver of the embryo into the throbbing heart of a newborn child.

15

DNA'S

UNBROKEN TEXT

AMERICANS are generally inept at foreign languages, but they have a particularly hard time mastering Chinese and Russian. Many Chinese say they feel the same about English, and even the most nimble European polyglot may be stymied by the intricacies of Navajo. But perhaps no language is more complex or more daunting than the language of the DNA molecule: the genetic instructions inscribed in every cell that tell a young body how to grow, an older body how to survive, and a fertile body how to reproduce. Now, in an effort to decipher the great helical string of biochemical letters that make up the book of life, a handful of imaginative biologists are applying the techniques of linguistics to the study of DNA.

They approach genetic sequences as though they were reading lengthy passages written in an archaic and largely unfamiliar tongue, borrowing methods from the linguist's tool kit to find a bit of order amid apparent biochemical babble.

Assisted by advanced computer programs, researchers pore over strings of nucleotides — the subunits of DNA that are the molecular equivalent of letters — in search of patterns that will tell them where to find the "words," the critical regions of DNA that instruct the cell how to make proteins, how to cleave down the middle, or otherwise comport itself.

The words being analyzed are very short segments of DNA, about three to five nucleotides long, but the patterns are distinctive enough to help reveal where the most cardinal genetic information is buried, like the commands for amino acids from which proteins are built.

Dr. Edward Trifonov, of the Weizmann Institute in Rehovot, Israel, who has taken the specialty further than anybody else, compares genetic sequences to ancient tongues like Hebrew, Etruscan, and Latin. In those languages, texts were written in an unbroken fashion, with no spaces separating one word from the next. Similarly, the DNA molecule sitting in its densely packed configuration in the cell is constructed of an unbroken string of billions of nucleotides, with no pauses between the end of one word indicating a unit of a protein and the start of another amino acid building block. Among the ancients, writing was the privilege of the upper classes, who felt no need to put blanks between their words; the DNA molecule apparently has the same sense of elitist privilege, and whoever doesn't get it can get out. But even though the run-on nature of genetic information complicates the task of decoding the sequences, Dr. Trifonov and others have spotted a number of the literary devices that DNA prefers. They are learning how to discriminate between those passages of the molecule that rank as genuine words and the much longer stretches of nucleotides that seem not to encode useful information and are often dismissed as junk DNA. They find that the pithiest regions are those with the most varied phrasing, while the intervening and least enlightening sequences are as random and repetitive as a series of keys typed by an infant.

Through this sort of approach, genetic sequences can be categorized as commands for amino acids, commands for turning genes on and off, and instructions for making the DNA fit neatly into the cell — much as linguists classify words as nouns, verbs, and modifiers. One can scroll through a modest sample of DNA from any organism and quickly identify the creature of origin or even find a molecular version of curses: groupings of nucleotides that are only rarely encountered because they seem to jeopardize the structural integrity of the DNA molecule.

Dr. Trifonov and and his colleagues have gone so far as to give the language of the genes a name and to compile a dictionary of significant genetic sequences and their meanings. They decided to call the language "gnomic," for reasons both staid and playful. For one thing, it is the language of the genome, the entire package of genetic material harbored in a cell's nucleus — drop the first *e*, and you get *gnome*. The name also evokes the gnomes of myth: the small, misshapen beings who dwelled in the earth, guarded its treasures, and wrote with silver pens, by moonlight, in secret scripts. Finally, a gnome is an apothegm, a concise expression of a general truth, and the genetic code is nothing if not a general truth so succinct that, with a mere four letters, it can speak of all life that has ever crawled or fluttered or sprouted or lumbered across the surface of blue planet Earth.

The theories of biolinguistics are a subset of a larger science, computational molecular biology, one of the most fashionable disciplines in biological research. Scientists are amassing so much information about genes and genetic sequences that it is only through the use of a framework like linguistics that they can interpret the incoming rush of data. In computational molecular biology, researchers search through computer databases of all known genetic sequences for even the vaguest similarity between one string of nucleotides and another. When they detect a relationship, they can design an experiment to determine whether the likeness is meaningful, implying some evolutionary commonality between two distinct genes, or is entirely coincidental.

One spur behind computational biology is the Human Genome Project, the international, multibillion-dollar venture to spell out all three billion nucleotides that constitute our DNA. The project involves sequencing huge stretches of genetic material on the chromosomes in an almost mechanistic fashion, throwing the information about the millions upon millions of nucleotides into a computer for deeper interpretation at a later date. Linguistic analysis is one possible approach to a thoughtful interpretation of what our genetic legacy may be.

The idea of thinking of genes as a language is not really new. The science of molecular biology first burst to life in the 1940s, which happened to be the time when social scientists were exploring the nature of communication and language. Hearing all the excited chatter about linguistics, most biologists were ready to think of the genome as a communications system. However, the power to turn armchair maunderings into plausible results awaited the advent of the genetic engineering techniques, by which the biochemical components of genetic sequences can be tweezed apart and spelled out. Biologists also needed to compile enough information about the genetic sequences from a variety of organisms before they could make the comparisons necessary to show genetic patterns.

As communications systems go, DNA is at once no frills and all frills. It's gnomically constructed of four nucleotides — cytosine, adenine, guanine, and thymine — which, when pieced together in varying order by the thousands, hold the instructions for the making of any protein. At the same time, only a tiny fraction of our DNA appears to be devoted to telling the cell how to survive and reproduce. Of the three billion nucleotides in our genome, a mere ninety million are thought to be constituents of our 100,000 genes. The remaining 97 percent of human genetic material, the streams and streams of Cs, As, Gs, and Ts found up and down the double helix, is either innocuous filler — so much molecular polystyrene to pack the genes safely — or serves a deeper purpose that has yet to be determined.

Above all, biolinguistics provides a method for researchers to pick out the core 3 percent from the biochemical background noise. They are trying to spot those words without having to worry about what the words say. Dr. Trifonov has devised a computer algorithm that can pick out a meaningful component from a long genetic sequence by searching for what is called a contrast word, a series of nucleotides that always or almost always appear in the same order. Such a sequence is likely to be an internal component of a real word, not neighboring parts of adjoining words or an entirely random set of letters. For example, an exhaustive analysis of English might reveal the contrast word *ookie,* the core of the word *cookies.* If you come on that sequence, Dr. Trifonov says, you know you'll find the letter *c* upstream, and the letter *s* downstream.

In the language of DNA, the contrast words are sequences of about five nucleotides that are found in a given order too often to be random, which suggests that they are words. That is, they have a message to convey; they are not the hems and haws of supposed junk DNA. The message may be a signal for creating one or more amino acids, or it may be a message for making parts of transfer RNA, little molecules that help to turn a DNA sequence into a working protein. Or the message may be something entirely unknown that deserves further exploration. As a rule, the portions of the genome with the greatest complexity and richness of contrast words are areas that instruct the cell on some aspect of survival. Moreover, within those complex regions are different dialects. Some parts of the genetic molecule consist of sequences that clearly are messages for building proteins — say, the message for synthesizing a growth hormone or an enzyme that helps metabolize food. Other sequences may be more architectural in nature, like the dotted lines on a cut-out doll telling you to fold here or insert tab there. In this case, the sequences serve as instructions on how the DNA molecule should be bent and kinked to assume its proper conformation within the cell nucleus.

Scientists have found words within genomes that are exceedingly rare, as though the sequences were as distasteful to the organism as profanities. Viruses that invade bacteria, for example, have few genetic sequences that bacterial enzymes can latch on to, limiting the bacterium's ability to destroy the viral DNA. These sequences cannot be entirely avoided, and bacterial enzymes do spy an occasional target on viral DNA to attack, but natural selection is a rigorous censor and keeps these dirty words at an absolute minimum.

Yet with all their efforts to understand the linguistics of DNA, biologists finally must appreciate that the idiom of the genome far surpasses the complexity of any language we have yet devised. Some gene sequences can be thought of as carrying multiple messages, one nested within another like a biochemical triple- or quadruple-entendre. For example, the phrase that tells the cell how to make a protein actually carries three distinct sets of instructions. First, it teaches the cell which amino acids make up the protein. Second, it informs the cell how to create the intermediary chemical message of RNA needed to construct a protein. Third, it tells the cell how to bend the newborn protein into its final operative shape. And it sends all three messages simultaneously.

The genome as a whole may be thought of as a literary masterpiece open to a multitude of interpretations. After all, in every one of us, each cell bears the same set of instructions — the same DNA poem, the same 100,000 genes, the same padded language between. Yet the cells of our liver look and behave very differently from those of our bones or brain. The text they're reading is so subtle, so richly layered with nuance, that each can find within the words the story of its own life.

III

SLITHERING

16

ADMIRERS

OF THE SCORPION

TO THE ANCIENT CHINESE, snakes embodied both good and evil, but scorpions symbolized pure wickedness. To the Persians, scorpions were the devil's minions, sent to destroy all life by attacking the testicles of the sacred bull whose blood should have fertilized the universe. In the Old Testament, the Hebrew King Rehoboam threatened to chastise his people, not with ordinary whips, but with scorpions — dread scourges that sting like a scorpion's tail. The Greeks blamed a scorpion for killing Orion, a lusty giant and celebrated hunter.

Throughout history and across almost every cultural boundary, scorpions have had a rotten reputation. And if the truth be known, they deserve it. They're nasty and they're not afraid of anybody. They can kill you or throw you into a seizure. Even those species whose venom is relatively innocuous can deliver stings of incomparable pain, "like flaming bullets twisting inside you," as a victim once put it.

Yet malignancy has its magnificence, which is why an elite and growing cadre of researchers are dedicating their careers — and braving nature's version of Uzi fire — to the study of the strange lives, violent nights, and brutal loves of scorpions, the nocturnal relatives of spiders and other eight-legged creepers known collectively as the arachnids. Long neglected in favor of spiders or their distant six-legged cousins, the insects, scorpions are finally winning scientific attention and respect.

They deserve that, too. If somebody were to put together a *Guinness* of invertebrates, scorpions would merit multiple entries. They are some of the biggest, meanest, longest-lived, most sensitive, most maternal, least fraternal, slowest, quickest, and most luminous creatures among the arachnids and insects. They are perhaps the oldest terrestrial animal on the planet, yet they have features that make them seem like thoroughly modern mammals. Biologists from Frankfurt, for example, have discovered that one of the largest species of scorpion, found on the Ivory Coast of Africa, is social to a degree unheard of among the normally solitary arachnids. Males and females, which can weigh almost three ounces apiece and measure up to eight inches in length, live together and rear their young for two years or longer. In caring for their offspring, the adults will kill rodents, frogs, and other vertebrates, strip the prey apart, grind it up, and feed the predigested stew to their young.

But scorpions are not always model spouses and parents. Some species are aggressive cannibals, deriving 25 percent of their energy by consuming their neighbors, their mates, their own young. In areas where more than one species compete for resources, the scorpions engage in elaborate interspecies feasting that would make the Borgias look like the Brady Bunch, with the elder members of the smaller species eating the offspring of the bigger species, the bigger species devouring the more diminutive adult scorpions, and two adults of similar size clashing for the right to remain, however fleetingly, on top of the food chain.

Researchers have gained new insights into scorpion mating, among the least romantic affairs in nature. Males and females engage in lengthy and violent waltzes, moving to and fro, to and fro, front legs gripping front legs, mouth parts locked together, and tails whipping forward, as the male repeatedly stings the female and the female thrashes about, as though furious at being dragged around. Sometimes, after copulation is completed, the female, who is almost always heavier than her mate, will exact revenge for the ordeal by consuming her partner. Not that she has any right to be so annoyed: the extended dance evolved in part to allow the female to assess the male's strength, heft, and genetic worth before accepting his sperm.

Naturalists from Aristotle on have been bewitched by scorpions, but it was only with the introduction in the 1970s of a simple device, the portable ultraviolet light, that anything substantive could be learned about these little princes of darkness. Another of the scorpion's exceptional features is the ability to glow under ultraviolet light, like a psychedelic poster.

The exoskeleton of the scorpion is made of a tough layer of tissue that feels like fingernail but is composed of another type of cuticle protein, chitin. This coat reflects the ultraviolet rays from moonlight and other light sources so brightly that even a black scorpion will be a fluorescent shade of green or pink. Fossilized scorpions from 300 million years ago still gleam brilliantly under ultraviolet light. The glow may have evolved to attract insects, which are drawn to ultraviolet light, or it may be an incidental byproduct of the chitin's chemical nature.

Whatever the reason, the unmistakable shine, visible from twenty feet away, makes it easy to spot scorpions at night, when they emerge to eat, mate, fight, swing their stingers, or simply vegetate in the open air. They can be located, captured, marked, released, and recaptured for measurements of their metabolic rate, oxygen consumption, and the like.

Through these investigations, arachnologists learned that scorpions have changed little since the Silurian epoch, 400 million years ago, when they were pioneers in the quantum creep from sea to land. Once grounded, they dispersed widely, and although they are commonly associated with the desert, in fact the fifteen hundred known species occupy every ecological niche and cranny: rain forests, temperate forests, savannahs, grasslands, Los Angeles. Blind ones creep around caves half a mile underground; tiny ones burrow in the cracks of pineapples; stout ones cling to the slopes of the Himalayas fourteen thousand feet above sea level. About a thousand more species are thought to be out there, tails poised, awaiting their lucky discoverers.

All known species are carnivorous and are equipped with venomous stingers, but only twenty-five species pack enough toxin to kill a human being. The venom is carried in a gland on the back of the tail, which the animal can whip forward in a fraction of a second to sting a victim, sometimes repeatedly. The venomous brew comprises up to thirty neurotoxins, each designed to fell a different type of prey. Some of the neurotoxins have been found to be most effective against insects; others are best at paralyzing frogs and other small vertebrates.

Once a prey has been knocked out, the scorpion begins the lengthy business of liquefying the victim. Like spiders, scorpions digest their food before consuming it, spitting out enzymes to dissolve the prey into a broth that it can suck into its mouth. Scorpions have other things in common with spiders. When they live in the same neighborhood, the two arachnids compete for the same resources, the insects. Yet scorpions have a distinct advantage over their competitors: that is, a taste for them. And given their usually superior size, scorpians can often turn their rivals into prey with little fear of reprisal. In areas where scorpions abound, spider populations are generally kept in check.

But a scorpion is by no means immune to predation. Although it can thwart some potential attackers with its venom, it is so meaty that owls, bats, snakes, and other animals will endure the sting for the sake of a hearty meal. Assuming they avoid being consumed, scorpions have the potential to live fifteen to twenty-five years and perhaps beyond, longer than any other known arachnid or insect. Contributing to its longevity is the scorpion's miserly metabolic rate, which is slower than that of any other invertebrate, roughly the equivalent of a growing radish root. Creatures with slow metabolisms generally live far longer than those which burn energy at a rapid clip, as most small animals do.

The scorpion's sluggish metabolism protects it from extremely harsh conditions of heat and cold on virtually no food or water. It can live for more than a year without eating, and its covering of a slick of wax seals in water. Even in urinating or defecating, it conserves water, releasing nothing but a powder of waste products.

Everything about the scorpion turns out to be extended in time. It takes close to seven years to mature and gestates its young for up to a year and a half, a pregnancy rivaled only by the elephant's. Scorpion mothers even have something like a mammalian placenta, which nourishes the young internally, another feature unique among invertebrates. The offspring are born live and crawl onto the mother's back for another two to six weeks of external development.

Those in the scorpion business say they are most impressed by the animal's exquisite sensitivity. Whereas other arachnids and insects have nerve cells disseminated up and down their body, a scorpion has a cluster of neurons in its head, lending it brainlike processing powers. It navigates by starlight and is practically an ambulatory seismograph. Slit-like organs on the creature's eight legs can sense surface disturbances from an insect walking on sand as much as three feet away. It can further extend its menu with flying insects. As the prey approaches, it lifts its front pincers in the air, and ultrasensitive hairs on the little claws begin to vibrate. The speed and direction of these quiverings tell the scorpion when the time has come to make its snatch.

Sensitive though they must be to find their food, scorpions also rely on senses tuned to mating and escaping. In recent studies of male scorpions, biologists determined that two strips of sense organs running down the middle of the animal's chest, called pectens, can sense seductive pheromones left behind by just a single foot of a female. The pectens also help the male find a vital stage prop for the mating dance: a stick on which he can deposit his sperm packet. During the dance, he must drag the female over to the stick, release his sperm, and help position her atop the stick. Eventually, the female will open her genital slit, located between her own pectens, and suck up the sperm packet. The male's pectens also tell him, through chemical signals, when a female has completed intercourse and is on the verge of attacking him for her postclimax snack. He will immediately try to break away from her and escape, but about 10 to 20 percent of the time he fails and is eaten.

Indeed, most scorpions are so notoriously hostile toward one another that some scorpion specialists would like to learn more about the handful of giant species that live in relative harmony in colonies, resisting even the temptation to swallow their tribe. The cooperative species may have some kind of colonizing pheromone that tells them not to attack each other and to be more social, and such a tranquilizing chemical might be useful to know about.

The notion of a cooperation pheromone is sweet, but even if scientists manage to isolate the substance, it can only be hoped that the scorpion retains its reputation as a ruthless night fighter, irrepressible cannibal, sexual athlete, and devil's handmaid. If nothing else, the scorpion, as the Methuselah of invertebrates, just may outlast any efforts to tidy up its delicious bad name.

17

PARASITES

AND SEX

To the ancient greeks, the word *parasite* meant one who ate at the table of another. Far from having the decency to sit down for dinner, most parasites suck the blood, sip at the gastric juices of the intestines, pierce the nourishing warmth of muscle tissue, and otherwise leech rudely off the fluids and labors of their unwilling hosts. Yet for all their repulsive traits, worms, mites, fungi, viruses, and a rogues' gallery of other parasitic organisms that derive their nutrients from a larger species merit a few moments of regard. They may freeload, they may try to keep a low profile while draining you dry, they may be tiny enough to skulk inside your cells, but in the grand theater of evolution, they are giants.

Many of the outstanding features we see in a broad spectrum of animals and plants evolved to counter the relentless pressure of parasites ready to colonize and exploit their every square picometer. Parasites are part of the reason that we and most species have sex, rather than merely cloning ourselves in twain: we must stir our genes together in ever-changing combinations to develop resistance to parasites. The need to evade parasites may have been the force driving some birds, fish, and mammals to become migratory or to spend part of every year in isolation from their potentially pest-ridden fellows.

And while it is natural to identify with beleaguered hosts everywhere, consider life from a parasite's-eye view. Seeking to understand why many species of parasites go through multistage life cycles, passing from one host to another, scientists have unearthed macabre examples of host-parasite relationships. For example, there are two types of closely related parasitic worms, each able to influence the behavior of mice to suit their own needs. One worm will prompt a mouse to become hyperactive, dashing through fields so frenetically that it attracts the attention of a predatory bird. On ingesting the mouse, the bird consumes the worm as well, thereby providing the necessary next home for the parasite's larvae. The related worm, on the other hand, will cause a mouse to become sluggish, heightening the chance that it will be easily stalked down by the carnivorous mammals this worm prefers for its second shelter.

Other parasitic larvae have been found to drive a host snail mad, forcing the creature to make a suicidal ascent to the top of a blade of grass instead of hiding underneath the foliage. A few of the invading larvae migrate to the snail's antennae, turn vivid colors, and pulsate, thereby transforming the hapless gastropod's feelers into a reasonable facsimile of a caterpillar, which catches the attention of a bird. Once inside the guts of the bird, the larval worms can mature and reproduce.

The bright new era in parasitology stems partly from the extraordinary advances in the study of the human immune system. As investigators deciphered its complexity, they came to consider the varied spectrum of parasites and pathogens that the immune system evolved to attack — with varying levels of success. All told, the impact of parasites on human evolution and human affairs has been stupendous. Although most parasitic diseases are now rare among those in developed nations, the majority of the world's people are hobbled by one or more types of parasite. By some estimates, the amount of human blood sucked by hookworms in a single day is equivalent to the total blood of about 1.5 million people. About half of all humans who ever lived have died from malaria, which is caused by the mighty

Plasmodium, a protozoan parasite. The Roman Empire was undermined by malaria. The early American colony of Jamestown had to be established three times because of malaria.

Nobody has a firm idea of how many parasite species there are, or even what constitutes a parasite. By the generally accepted definition, a parasite must derive most or all of its nutrients and resources from another animal or plant species, and it must be smaller than its host. But while many parasites, like viruses and bacteria, are microscopic or nearly so, some types of worms reach three feet or more as adults. Parasites frequently are harmful to their hosts, although the degree of virulence varies widely. Some, like many viruses, sicken and kill the animals they infect; others cause only the mildest of malaise; still others are almost completely innocuous. Every so often, a host figures out a way to put its parasites to work, at which point the spongers become symbionts. The bacteria that live in our gut and help digest our food offer a familiar example of welcome infestation.

Parasitism is such an appealing way to earn a living that the majority of the earth's organisms have adopted it. A number of parasites, like ticks, are generalists, hopping readily from one warm-blooded creature to another. Many more are remarkably specific. There are mites that can survive only in the rectum of a giant tortoise, worms that fit snugly into the quills of a single species of bird, and mites that live exclusively and harmlessly at the base of human eyelashes. Most parasites are themselves burdened with parasites, which prompted the novelist and satirist Jonathan Swift to write in his 1733 work *On Poetry:* "So, naturalists observe, a flea hath smaller fleas that on him prey. And these have smaller still to bite 'em; and so proceed *ad infinitum.*"

Until recently, parasitologists studied their subjects largely in an effort to eradicate them, but lately they have turned to parasites for knowledge about profound evolutionary mysteries, one of the biggest being why sex evolved. On the face of it, sexual reproduction is cumbersome and irrational, far less efficient than propagation by the neat clonal copying of the mother organism. Some simple animals and plants do procreate asexually, but most higher species opt for the Noah's ark approach to life, bounding up the on-ramp male by female. And, sublime though sex may be at its best, one cannot help pondering its ubiquity. Some biologists have suggested that organisms must create diversity in future generations to guarantee at least a few offspring who will survive changing climate, uncertain food sources, and other environmental fluxes. Another theory has it that if babies differ sufficiently from one another, they will be less likely to compete directly for resources.

But by far the greatest evidence for the evolution of sex comes from the study of parasites. Varying one's offspring through sexual reproduction turns out to be an ideal method for outwitting parasites, which prefer to infest creatures that are similar to hosts they have exploited before. The benefits of sex for thwarting parasites can be seen in the unusual life cycle of an aquatic snail in New Zealand, *Potamopyrgus antipodarum*. Some of the female snails reproduce sexually, others asexually, and their decision on which switch to hit has nothing to do with libido. Researchers studying sixty-five populations of the snail found a strong correlation between sex and the prevalence of the most deleterious snail pest, a type of nematode. In lakes with a light load of the parasitic worms, there were few male snails, and the females reproduced asexually. In lakes awash with nematodes, male snails abounded as well as females, suggesting that the females required a male for reproduction and consorts to help them spawn more resistant babies.

Another discovery reveals that some parasites consider sexual reproduction in their hosts to be such an anathema that they will strenuously suppress it. Among species of grasses that can reproduce either sexually by producing flowers or asexually by sending out carbon-copy shoots, the fungus that parasitizes them wants no part of genetic variability. On infecting a plant, the fungus heads straight for the grass's sex cells and destroys them, but does no further damage to the host. Thereafter, the infected grass can produce only shoots, each of the offspring genetically identical with its forebears and thus vulnerable to fungal infiltration.

In a story not unlike that of *Invasion of the Body Snatchers,* a fungus that infects a flower related to the carnation not only sterilizes the plant, it transforms the blossom into a fungal factory. In the stamen of the flowers, where the pollen of the plant normally would be found, an infected plant displays a bristle of fungal spores. Pushing perversity to the extreme, the fungus causes its floral hosts to grow bigger and showier flowers than those of normal plants, attracting pollinating insects and ensuring the transmission of parasitic spores.

Parasites attack the mobile as well as the grounded, and in retaliation the mobile become more peripatetic still. The red spotted newt found in mountaintop ponds of western Virginia, for example, carries a parasite related to the agent of deadly African sleeping sickness in humans. However, the parasite appears to do the newt little harm, a benignity that happens to be the result of the amphibian's migratory habits. At the time when the newt harboring a more virulent strain of the parasite might transmit the pathogen to fellow newts, the animals are spending months roaming alone through the woods, rather than congregating in ponds. Those newts carrying a malevolent parasite die off during their migrations, leaving only the amphibians with a mild strain of the parasite, and they return to the pond to mate.

In like manner, birds that fly each year from North to South America may be avoiding more than bad weather. During the nine months down south, the animals do not breed and are not particularly close to one another, limiting the chance for pests to feather-hop.

Nor is life easy even for a parasite that has successfully infected a host. Because the parasite is utterly helpless without the host, the latter's death can often mean termination for the parasite, too. Thus, parasites have expended a great deal of energy evolving methods to secure their transmission to another host, like making a carrier sneeze, or changing the host's behavior sufficiently to warrant its getting eaten by Host Number 2.

But some parasites go further, demonstrating a spirit of self-sacrifice in their quest to keep their fellows alive. In my favorite parasite parable, a liver fluke known as a lancet begins life as an egg laid in the intestines of a sheep. It and the other eggs are expelled when the sheep defecates and are then eaten by land snails that feed on sheep feces. Inside the snails the eggs hatch and develop into larvae, which are again voided by the host, this time wrapped in a slimy packet that ants find irresistible. Once ingested by ants, the lancet larvae divide and conquer, some moving into the ant's intestines, where they develop into a new, infectious stage, and a few invading the ant's brain. These brain-worms so disturb the ant that, early in the morning and late in the evening, it does something no sane ant would do: it climbs to the tip of a blade of grass just at the moment when sheep will be grazing. Safely inside the belly of a sheep, the larvae from the ant's abdomen mature, mate, and lay eggs, starting the entire rococo cycle over again.

As for the brainworms, they sacrificed themselves for their kin. They didn't become infective and they didn't reproduce, but they died so that others could thrive. If there is such a thing as altruism, here's an excellent example; but from the ant's perspective, at least, Mother Teresa this fluke is not.

18

THE SCARAB,
PEERLESS RECYCLER

IN THE VAST WORLD of beetles, they have the stamp of nobility, their heads a diadem of horny spikes, their bodies sheathed in glittering mail of bronze or emerald or cobalt blue. They are symbols of rebirth, good fortune, the triumph of sun over darkness. The ancient Egyptians so worshiped the creatures that when a pharaoh died, his heart was carved out and replaced with a stone rendering of the sacred beetle.

But perhaps the most majestic thing about the insects known romantically as scarabs and more descriptively as dung beetles is what they are willing and even delighted to do for a living. Dung beetles venture where many beasts refuse to tread; they descend on the waste matter of their fellow animals and swiftly bury it underground, where it then serves as a rich and leisurely meal for themselves or their offspring. Each day, dung beetles living in the cattle ranches of Texas, the plains of Africa, the deserts of India, the meadows of the Himalayas, the dense undergrowth of the Amazon — any place where dirt and dung come together — assiduously clear away millions of tons of droppings, the great bulk of it from messy mammals like cows, horses, elephants, monkeys, and humans.

We are all beholden to dung beetles, nature's original recyclers, without which our planet would be beyond the help of even the most generous Superfund cleanup project. Inspiring the insects to take on their unenviable task with such dispatch are powerful market forces, the ferocious interbeetle competition that erupts each time a mammal deposits its droppings on the ground. Every dung pat is a complex microcosm unto itself, a teeming habitat not unlike a patch of wetland or the decaying trunk of an old redwood, although in this case the habitat is thankfully short-lived. For scarabs, it may be said that waste makes haste, and tens of thousands of dung beetles representing as many as 161 different species will converge on a single large pat of dung as soon as it is laid, whisking it away within a matter of hours or even minutes.

The diversity of beetles that flock to a lone meadow muffin far exceeds what anybody would have predicted as likely or even possible, and scientists lately have begun to rethink a few pet notions about how animals compete for limited goods and what makes for success or failure in an unstable profession like waste management. They are learning that beetles evolved a wide assortment of strategies to get as much dung as possible as quickly as possible, to manipulate it for the good of themselves and their offspring, and to keep others from snatching away their hard-earned spoils.

Chance and circumstance also play a central role in determining who reaches a prized resource first and who is able to make the most of it. The knowledge gleaned about the dung beetle community enlarges scientists' understanding of how species compete for more conventional resources, including plants or prey animals. Many of the most piquant findings have been gathered into *Dung Beetle Ecology*, edited by Ilkka Hanski and Yves Cambefort. This book, occasionally technical and graph-dependent, nevertheless manages to transform a beetle one previously might have preferred not to dwell on at all into an insect of such worthiness, respectability, and charm that one only wishes there were a bulk-order catalogue from which a few hundred thousand beetles could be purchased for distribution as needed — in the local dog-walkers' park, say, or down at City Hall.

Dung beetles are among humanity's greatest benefactors on many accounts. Not only do they remove dung from sight, smell, and footstep, but by burying whatever they do not immediately eat they add to the soil fertilizing nitrogen that otherwise might be lost to the atmosphere. Like earthworms, the beetles churn up and aerate the ground, making it more suitable for plant life. Their larvae consume parasitic worms and maggots found in dung, thus helping to cut back on micro-organisms that spread disease.

As beetles go, scarabs are exceptionally sophisticated. In Africa and South America, where some species are the size of apricots, the beetles may couple up like birds to start a family, digging elaborate subterranean nests and provisioning them with dung balls that will serve as food and protection for their young. Nor are these dung balls, called brood balls, just slapdash little mudpies. With a geometric artistry befitting the sculptor Jean Arp, the beetles use their legs and mouth parts to fashion freshly laid dung into huge spherical or pear-shaped objects that may be hundreds of times the girth of their creators. Some beetles even coat the balls with clay, producing orbs so large, round, and firm that they look machine-made. In fact, people who have dug up these creations have mistaken them for cannonballs.

Still working as a duo, the beetles roll each ball away from the dung pat and down into the underground nest. The female lays a single egg in each brood ball; among the largest species there may be only one ball and thus one baby per couple, a surprising show of restraint for a member of the class *Insecta,* which is traditionally known for its emphasis on fecundity.

Safe within its round cocoon, the larva feasts on the fecal matter. As the infant develops over a period of months, the mother stays nearby and tends to the brood balls with exquisite care, cleaning away poisonous molds and fungi to ensure that her young will survive to emerge from its incubator as an adult.

Other scarabs are superspecialists, their habits streamlined to harvest the ordure of one type of mammal alone. Such beetles may cling to the rump fur of a kangaroo, a wallaby, or a sloth, awaiting the moment when the final stage of mammalian digestion is complete and then leaping on the droppings in midair. Some beetles dine solely on giraffe waste, others on the excretions of wild pigs. Panamanian beetles fly each morning up to the treetop canopies where howler monkeys sleep. They wait for the primates to awake and do their morning business, then quickly latch on to the released flotsam and sail with it a hundred feet to the ground, where they can bury it.

The majority of dung beetles, however, are generalists, able to make a meal and outfit a nest with any droppings they find. Of keenest interest to the scarabs are the generous patties provided by large herbivores, which by the nature of their digestive system must void themselves frequently. The average cow produces ten to fifteen large pats per day. Elephants offer about four pounds of waste every hour or so, and an elephant pat can become a pulsating Manhattan of beetles, with different species exhibiting a splendid variety of tactics. Big scarabs roll huge balls of it to their nests several yards away, sometimes pushing the balls over logs and boulders; smaller dung rollers trundle off with more modest portions. A class of beetles called tunnelers bury big hunks of dung right beneath the pat; other beetles, pin-sized, live within the pat, munching on it even after it has begun to dry up and be of little use to the larger, more aggressive rollers and tunnelers. Robber beetles sneak in and try to steal balls painstakingly shaped by others. Joining the fray are many species of dung-eating flies. The scene resembles a fast-food outlet at lunchtime, with all the patrons grabbing something to bring back to their desks. A dung heap is also a sort of singles bar, where beetles in search of mates can meet and begin the joint effort of gathering the goods for their nest. Some of the larger beetles use dung in their courtship dances, the male lifting a deftly rolled bit of dropping and shaking it provocatively in a female's face.

All of which means that little dung goes to waste. One research team in Africa reported counting sixteen thousand beetles on a single elephant dung pat; when the scientists returned two hours later, the pat had disappeared. The incentive to move quickly is great. Not only does every beetle want to get away with the biggest slice of the pie, but while they are scavenging in an exposed heap of dung they are tempting to many insectivores. Birds, mongooses, monkeys, and other small animals can often be seen picking around in a herbivore's droppings, and it is not because any of them have turned coprophage. Some beetles protect themselves by evolving persuasive disguises. One scarab that frequents elephant dung, for example, resembles an undigested stick.

Behind the diversity of dung beetles is the resource they live on. Hard though it is to fathom, dung has much to recommend it. Most mammals digest only a fraction of the food they eat, and whatever they discard is rich in proteins, nutrients, bacteria, yeast, and other sources of nourishment. Best of all, dung is easy. Animals fight back against would-be predators, and plants generate poisons to ward off herbivores, but dung does not bother defending itself against consumption. It is the line of least resistance.

Dung beetles may prefer the droppings of the biggest mammals, but the insects originated more than 350 million years ago, before the appearance of such animals. Scientists speculate that dung beetles fed on dinosaur waste, but no beetle fossil has ever been found in the midst of petrified dinosaur feces. With the rise and spread of the great mammals around the world, dung beetles likewise began to diversify and multiply. Indeed, the two events occurred in parallel, and some researchers have suggested that large mammals might never have reached the population densities seen in places like the African savannah without the aid of beetles to clean up their waste, thus allowing the plants they feed on to keep growing. They are key organisms in the environment.

From the dawn of agriculture and the domestication of waste-heavy livestock animals, human beings also have recognized the value of the beetles, the ancient Egyptians taking their admiration to the greatest lengths. One researcher suggests that the Egyptians' tradition of mummifying their kings and burying them in pyramids was modeled after the burial of a beetle larva in a dung ball. Just as a beetle rises from dirt to a new life, so, the Egyptians believed, their pharaoh would be reborn from his interred cocoon. The Great Pyramids of Giza may thus be thought of as glorified dung heaps.

The benefits of scarab beetles have not gone unnoticed in our own time. *Dung Beetle Ecology* recounts the ambitious and largely successful effort by the Australian government to import thousands of exotic dung beetles to reduce the mountains of dung generated by cattle and sheep. Those livestock animals had been brought to the continent within the preceding two centuries, and indigenous Australian dung beetles, accustomed to the bite-sized droppings of kangaroo and koala dung, were unable to handle the foreign animals' enormous output. By the 1960s, the fecal problem had reached crisis proportions, and the dread native bush flies, which breed in excrement, had achieved levels pestilent enough to give birth to the famed "Australian salute," a brush of the hand across the face to swipe away flies. But with the introduction of two dozen species of beetles from Asia, Europe, and Africa, the dung problem has begun to subside. In recent years, many parts of southern and western Australia have become almost entirely free of the dung-breeding flies, and pastures that once were coated with a carapace of cattle dung have been restored to useful verdancy.

On a more theoretical scale, ecologists have also learned much from the insects. Scientists historically believed that more than one species could not coexist in the same ecological niche without showing some differences in their use of resources. By a rule expressed in mathematical equations, it was asserted that one competitor must eventually prevail over the others.

But, given the diversity of dung beetles living on a single pat, nature obviously doesn't stick to the code. Entomologists investigating the beetles realize that dung as a resource has a few distinguishing characteristics. It is far more ephemeral than, say, a patch of flowers or a burrow of rodents, being here today and gone today. And its distribution in the environment is wholly random, with no easily defined rules about when or where it is likely to appear. To most animals, everywhere and anywhere is a potential lavatory.

Therefore a strong element of happenstance must be figured in to any calculation of the dynamics of the dung beetle community. As it turns out, randomness fosters the survival of many competing species. Some of the larger dung beetles may be inherently more gifted than others at monopolizing prodigious quantities of dung once they get to the pat; but because a smaller, weaker beetle is as likely to be close to the site of miraculous presentation as is a larger and more aggressive beetle, the weaker species will always have a shot at a food source, and the superior scarab will not be able to outcompete it every time. The randomness of the distribution of dung adds a crucial element of chance to survival, and that element incidentally favors the coexistence of many species. In nature's casino, fortune as well as fitness keeps the game alive.

19

THERE IS NOTHING

LIKE A ROACH

IF ABSENCE makes the heart grow fonder, then perhaps the moment has arrived to consider a modest celebration of the cockroach.

In recent times, many city dwellers have been able to stride into their kitchens at night with the newfound confidence that they can flick on the light, take a glass from the cupboard, even grab a few cookies from a box on the counter — all without the odious sight of dozens of greasy brown cockroaches scattering for cover. A new generation of insecticides, packed into discreet little disk-shaped bait traps called Combat or applied in more potent concentrations by professional exterminators, has helped bring the ubiquitous German cockroach to its six spindly knees.

The creature is far, far from nearing extinction, and remains a serious pest in restaurants, hospitals, and many inner-city housing projects. But the new insecticides, the amidinohydrazones, introduced in the mid-1980s, have made a significant dent in the less extreme infestations. There was a time when people traded cockroach war stories at parties; now everybody wonders how something as cute and moderne as Combat can do such a dirty job so well.

Entomologists calculate that the new chemicals will cut the German cockroach population by 50 percent to nearly 100 percent, depending on the severity of the infestation. Equally heartening, cockroach studies around the country indicate that the insect shows no signs of developing resistance to the amidino-hydrazones, as it has to nearly every other noxious compound leveled against it in the past. And should the creature somehow manage to mutate beyond the might of the current pesticides, other highly effective compounds are waiting in the wings, many of them based on a subtle understanding of the insect's biology and habits.

So, now that we no longer need share every meal and inch of shelf space with unwelcome squatters, we might consider something less than Armageddon in our approach to cockroaches. We may feel, if not outright affection, at least a detached admiration of their antiquity, persistence, and resourcefulness. Among species found in the tropics — where the creatures know their place and that place is not ours — the insects engage in the most unroachy sorts of behavior. Some female cockroaches are devoted mothers, carrying their offspring in little pouches, kangaroo-style, rather than simply dropping their eggs and leaving the nymphs to their own metamorphic devices, as most insects do. One type of cockroach even performs the insect equivalent of breast feeding.

Good parenting is not restricted to the mother. While many male animals contribute nothing to their progeny's welfare beyond a charitable donation of DNA, some male cockroaches carry paternal care to a far greater length, dining off bird droppings for the sole purpose of extracting nutritious nitrogen that they then bestow on their developing offspring. One variety of cockroach that lives in Central American tree bark is as social an insect as a termite or bee. The males and females couple up to nurture their nymphs for the five or six years it takes the species to reach adulthood. All members of the nest maintain a sense of group identity and cooperation through the use of mutual grooming, antenna stroking, and placating pheromones, chemical signals that are secreted by glands on the thorax of one insect and detected by the antennae of another.

Cockroaches are exquisitely sensitive to the slightest breeze or the faintest smell, a trait made possible by their unusually long antennae. Such chemical and tactile sensitivity, combined with a nervous system built of enormous cells, makes the cockroach an ideal experimental organism for the study of how nerve cells work. Cockroach receptors for detecting air movement or chemical signals are located on the outside of the body, where they can be readily analyzed. And the cockroach's head will live and respond for at least twelve hours after the animal has been decapitated. Among neurobiologists, cockroaches have become the insect version of the white rat, the subject of textbooks that describe how to manipulate and dissect them and why it's worth bothering. If nothing else, cockroach researchers need never fear that animal rights activists will break into their lab at midnight to liberate their experimental subjects.

One scientist at the University of Illinois has made it her mission to bolster the cockroach's reputation. Every year May Berenbaum holds an "insect fear film festival," using clips from movies to stimulate the public's interest in insects and to dispel myths. And one of the most commonly seen bit players is the cockroach. In both short films and animated features, the cockroach is often portrayed as a sympathetic character. One offering, *All's Quiet in Sparkle City,* an antiwar film from the early 1970s, equates efforts to eradicate cockroaches with genocide. A 1989 comedy, *Dr. Ded Bug,* is seen from the insect's perspective as a frenzied chef attempts to hunt down and kill a cockroach. Like Mickey Mouse, cartoon cockroaches talk in high, chipper voices and rarely stop smiling. In animated films, vermin are your friends.

Whether or not cockroaches become one's thorax buddies, the insects deserve respect for their advanced age and their diversity. Fossils of cockroach-like species have been found dating back 280 million years, and some entomologists estimate that the creatures may have originated in the Silurian period, 400 million years ago. By contrast, beetles are only about 150 million years old, and butterflies are a youthful 60 million years.

Cockroaches are found in nearly every part of the world, but the great majority of the four thousand known species live in the equatorial belt (and an estimated thousands more remain to be discovered in the tropics, where practically everything remains to be discovered). Cockroaches range in size from a quarter of an inch to the forbidding Megablatta of Central America, which in length and girth approaches the dimensions of a small rat. Universal to cockroaches are long, segmented antennae, a leathery pair of front wings that give some warm-weather species the ability to fly but otherwise are vestigial, and the famous cockroach head, which is tucked under and slightly pointed toward the rear. (Everybody who knows and despises cockroaches is familiar with the unmistakable cockroach profile, and if the head of a praying mantis reminds you of the head of a roach, that's because the two are closely related.)

Some smaller cockroaches are exquisitely colored; they may be deep crimson, spring green, a creamy white, or a pale toffee. The most dazzling specimen is sapphire blue with bronze flecks and slim red stripes, so pretty that one scientist suggested that if it were a bird, people would buy it and put it in a cage. But only one type of cockroach is frequently kept as a pet: the three-inch Madagascar hissing roach, which attempts to scare off predators by expelling a noisy blast of air through holes in its upper thorax. One reason the hissing roach makes a good pet is that it's covered with an armorlike cuticle and thus is appealing to hold and stroke. By contrast, most roaches are coated with a layer of oil that eases their passage into cracks thinner than a paper match.

The most physiologically elaborate of all cockroaches is the species *Diploptera punctata*. The female carries her embryos live, rather than in an egg case, and she is the only insect known to nourish her young in the uterus. The lining of the brood pouch, where about twelve baby cockroaches grow at a time, secretes a substance that some call cockroach milk. Like mammalian milk, it is rich in protein, carbohydrates, and fat. After the embryos grow a fully developed digestive tract, the mother produces the milk, and the embryos sit in the pouch ingesting the fluid through their mouths.

But while some cockroaches have evolved an extensive system of maternal care, others have opted for greater fecundity and the utmost behavioral flexibility. These are the species that have become pests to humans. Only twenty types of cockroach are classified as pests, and only two of these, the German and the American cockroaches, are broadly familiar, the three-inch American cockroach going by the name of palmetto bug in Florida and water bug in New York. These two pest species have fared so successfully in their strategy of taking up residence with humans that they no longer have an independent existence or any representatives of their kind in the wild. Whenever scientists in the field thought they'd stumbled on a free-ranging American or German roach, they always found a house somewhere nearby. The smaller German cockroach is an especially prolific breeder, able to spawn thirty or forty infant cockroaches every three weeks. If the population were left unchecked, a single female German cockroach could give rise to about forty million offspring in her two-year life span.

The insects grow to adulthood rapidly and molt frequently, which is why cockroaches can present a real health hazard for those with allergies. About fifteen million Americans suffer from cockroach allergies as their immune system mounts an overzealous defense against airborne particles of molted cockroach skin. The allergies often worsen with time and continued exposure, and entomologists who work with cockroaches say they develop wheeziness, skin rashes, and sinus trouble after years of pursuing their research.

For that reason, and because of the possibility that cockroaches transmit the unsavory microbes piggybacked upon them, even entomologists who like the insects in the wild spend part of their time devising better ways to thwart the pests. To determine where cockroaches congregate and why, one team of scientists designed an entire mock house. Its two hundred sensors monitored the microclimate every sixty-five milliseconds and at every possible spot — behind the walls, under the sink, up in the rafters. The high-wired house revealed that nothing will repel cockroaches as surely as good ventilation. The animals use almost indetectable air currents to sense the chemical signals of their mates, but any air movement approaching a draft will quickly and fatally desiccate the cockroach's coating. Thus, the best way to defeat cockroaches is by keeping the kitchen window open or putting a small fan in the cupboards or under the sink.

Another attack position calls for the use of the new generation of pesticides, which differ markedly from the old spray cans of Raid or Black Flag. Those pesticides were made predominantly from organophosphates and carbamates, potent neurotoxins that disrupt the transmission of impulses from one nerve cell to the next. But the poisons worked on only a single component of the nervous system, and some cockroaches had an inborn resistance to that method of assault. The resistant insects, of course, survived to propagate entire legions of resistant nymphs. The newer pesticides seem to be more global in their activity, affecting so many parts of cockroach physiology that it is unlikely that any one insect will have all the genetic traits necessary to withstand the storm. The active ingredient found in Combat, for example, interferes with multiple steps in the biochemical process by which a cell uses its stores of energy.

Of perhaps greater importance, the toxin works at attractively low concentrations. One reason cockroaches may have trouble developing resistance to the chemical is that even the few insects that survive a nibble of the poisoned bait become sterilized and fail to pass along their detoxifying ability to offspring, as survivors of the older generation of pesticides were able to do.

But though Combat works for now, keep in mind whom the odds really favor. Cockroaches have an astronomical reproductive rate and a fast life cycle. They've been living with us for thousands of years. They're patient. So while city dwellers may pray to God that the new pesticides retain their powers for decades to come, we have to wonder whether the divinity we're beseeching has a wise old human face and a long white beard, or whether its head is tucked under and pointing toward the rear.

20

THE PIT VIPER:
BIZARRE, GALLANT, VENOMOUS

STARING into the face of an angry rattlesnake is one of those things you are not supposed to do, and you know it. The thinking brain tells itself, "Why don't you get out of the way, fool?" while the primal brain resorts to a simple "Aaagh!" More unnerving still is the feel of the snake's rattle, a little castanet made of something like fingernail that vibrates ominously at fifty cycles per second. But this particular snake is safely harnessed in a laboratory rig that prevents it from striking or escaping, and the rig is held by Harry Greene, of the University of California at Berkeley, one of the world's authorities on rattlesnakes. So after I spend a few moments tentatively petting the reptile's tail and haunches, my visceral response subsides and the furious beauty of the creature shines forth. Its body is pure, pulsing muscle; its expression as proud as a prince's. I can't stop staring at the serpent's face: its flicking split ribbon of a tongue, its lidless yellow eyes. Where have I seen that face before? I reach for my apple.

Greene is chattering happily beside me, talking of the glories of snakes, of his plans to convert the public from fear to cheerleading, of his hope that someday the average American family will consider a visit to a timber rattlesnake den a wonderful way to spend their summer vacation. I assure him I'll speak with my travel agent.

Greene is a member of a small but passionate corps of herpetologists specializing in pit vipers, a group that includes diamondback rattlesnakes, cottonmouths, copperheads, sidewinders, and other commonly loathed snakes bearing on their face the two characteristic pits that explain the family name. Long ignored by the mainstream biology establishment, which takes mammals, birds, and even insects far more seriously than it does lizards too lowly to lift their bellies off the ground, pit viper science has lately come into its own. In the portrait emerging from this work, the snake is a lover at once gallant and violent; a killer whose brew of lethal and degradative toxins is among the most complex compounds in nature; and an evolutionary exemplar of streamlined design and elegant adaptations. Who but a snake could say, Look, Ma, no hands; look, Ma, no feet; look, Ma, no teeth, and say it with such feeling?

Behind every scientific fashion lies a touch of technical novelty, and the surge in serpentine research is no exception. Using recent advances in radio telemetry, biologists now outfit pit vipers with miniature radio tags and track each slippery and cryptic creature wherever it may be hiding. In the past, the same snake could almost never be found twice, but with telemetry, herpetologists have come to know individual snakes as intimately as Jane Goodall knows her chimpanzees.

Following prairie rattlesnakes in Wyoming, scientists found that, during mating season, males embark each morning from their den on a grueling, five-mile round-trip search for willing females. And as they travel, they crawl in a line so straight it could have been penned by a draftsman. Should they have to deviate from their path to scoot around a pond or boulder, they return to the straight and narrow as soon as possible.

As computer simulations demonstrate, there is a rationale for that obsessive-compulsive behavior. Female snakes tend to be distributed around the environment in random clusters as they seek a dinner of rodents, which themselves are clumped about. Male snakes covet females, and by following a straight path instead of zigging to one side or zagging to the other, the males optimize their chances of encountering mates. The straighter the path, the higher the possibility of their colliding with them.

And with that collision, the showmanship begins. A male snake rarely finds his loved one alone, but instead must engage in highly ritualized battles with competing males on the scene. The fights resemble nothing so much as disembodied arm-wrestling matches. Rearing up and entwining their necks like the two figures on a caduceus, the males attempt to force each other to the ground. They are gentlemen jousters, and they keep their venomous fangs tucked away, struggling for hours and often ending their encounter in a draw. But woe to the snake that cannot stand its ground. Among copperheads, a trounced male ends up so demoralized that for days afterward even far punier males, which normally would be loath to pick a fight, will take on the loser and defeat it. Behind the pitiable behavior of the defeated is a change in hormone levels, notably a rise in the stress hormone, cortisol, and a tumble in the testosterone concentrations that lend male snakes their randy aggression. On occasion, a female copperhead will take advantage of the loser syndrome. Approaching a potential mate, she will mimic another male, rearing up as though ready for battle. Should the mock display terrify the suitor, she will take it as evidence that the male is a loser and reject him as unfit for paternity. Females mate almost exclusively with winners.

But rarely. Most female vipers breed less than once every three to five years, which is why a fertile female is highly prized and mightily fought over. When an ovulating female emerges in spring, there may be hundreds of males waiting outside her den, a convocation called an explosive mating assemblage. When she appears, the frenzied male tournaments commence. After hours or days of fighting, there emerges a winner, who approaches the female and courts her with gentle chin rubs and tongue flicks. Eventually he may be awarded with copulation, an act that itself lasts for hours or days, as the male locks himself into the female with one of his two barbed and bifurcated genitals, called hemipenes.

The serpent's extraordinary powers are not confined to love-making. Its chemosensory skill as a hunter is thought to be the most acute in the animal kingdom. A viper's strike takes only a fraction of a second, but in that snatch of time the snake picks up the distinct chemical signature of its prey, essentially smelling the creature with its forked tongue. The odor molecules are conveyed from the tongue to a matched pair of sensory glands called the vomeronasal organ, located on the roof of the viper's mouth, where they create a chemical memory. The envenomed animal is allowed to wander off and die, but the viper can track it down wherever it may stumble. In laboratory experiments, researchers have watered down the scent of a prey rodent after the pit viper has tasted it, but they have not yet, after thousands of dilutions, succeeded in throwing the snake off the trail.

The venom of a pit viper also presents a challenge to toxicologists, who would like to understand it for the sake of improving antivenom treatments. About eight thousand people a year are bitten by pit vipers in the United States, the majority of them drunk young men who tease or maul snakes they find by the side of the road. Professional and recreational snake handlers in Western states consider it the essence of machismo to endure an occasional rattlesnake bite and often do not bother seeing a doctor afterward, with sometimes grim consequences. Even a relatively minor snakebite on a finger can result in permanent paralysis of the hand. More serious bites cause extensive destruction of tissue, internal bleeding, dangerously lowered blood pressure, and death. Harry Greene, who has not been bitten since he endured a tiny nip by a copperhead as a teenager, displays photographs in his lab of people who have been far more severely envenomed, showing legs, arms, and genitals that are masses of blackened, blistered, necrotizing flesh. Snakes normally will not bite you unless touched or seriously provoked, he says, but they must always be treated with caution and the utmost respect.

The reason for the deadly potency of the venom is the snake's need to fulfill multiple tasks with a single shot. Composed of dozens of neurotoxins, blood poisons, and degradative enzymes, the venom not only helps subdue and kill a prey animal, it also begins digesting the creature from inside out. Because a snake has no limbs to ferry its food back to a burrow for later consumption, it must consume what it captures at once, before another predator steps in to claim it. Lacking teeth to chew, a pit viper swallows its meal whole, moving its vertebral column over a creature that may be one and a half times its own body size. The snake then digests the prey internally over days or even weeks. Were it not for the degradative effects of the injected venom, the prey would putrefy within the snake's belly.

Pit vipers are by no means the gaudiest snakes — coral snakes, for example, are much brighter — but their camouflaging hues of gold, brown, mauve, black, and tan give them a lush, velvety appearance. Most of the 144 species of pit vipers live in North and South America, although several types are found in Asia. They thrive in the harshest deserts, the wettest Amazonian rain forest, and the gentlest meadows and grasslands. They live on the ground, underground, or in trees. They can be short and bulky or long and slim. During winter, they must hibernate, either singly or in great dens of hundreds or even thousands of entanglements, where they keep each other a degree or two warmer than they might otherwise be. In the past, such wintering dens were the targets of ranchers and farmers, who torched them or threw in explosives. Though the practice is now illegal in most states, some species of vipers have yet to recover from the extermination efforts.

Linking the pit vipers are two outstanding features: a shakable rattle or its nubby, silent precursor, and the deep facial pits. The pits are depressions in the bone covered by a thin membrane of sensory nerve cells that acts as a sort of lens to detect thermal radiation from the environment. The infrared signals, conducted from the membrane along nerve fibers back to the part of the brain that receives visual information from the viper's eyes, results in a thermal image that enhances the visual one.

However, traditional assumptions about the purpose of these

pits may be wrong. Researchers long believed that the sensory organs help snakes hunt their warm-blooded prey; but recent comparisons of the pit viper's habits with those of related pitless viper species suggest that the infrared detectors developed not for offense, but for defense. The pit viper uses thermal information from an approaching animal to determine whether the potential predator is bantam enough to be scared off with a threat, or whether the viper would do best to slink away. Harry Greene, the father of the new theory, points out that pit vipers and their pitless relations — which lack the benefits of infrared sensing — show no discrepancy in their hunting or eating habits. Both types of viper favor rodents and other small mammals, which they hunt by lying in ambush.

Where the serpent families differ is in their tactics for protecting themselves. Pitless vipers are designed for quick retreat. They generally sport a striped pattern, which creates an optical illusion that makes them difficult to see and catch. Nor do pitless vipers have the rattle to warn intruders that they will stand up for themselves. Pit vipers, however, will waggle their gourds or pound their nubby rattle precursors on the ground when disturbed. They usually are covered with splotches, which are of no visual aid in retreat.

There is some justification for the different approaches to danger. Pitless viper mothers do not bother defending their eggs, but instead seek a hidden spot where they deposit the eggs to hatch or be eaten. Pit vipers usually linger around, guarding their clutch for days to weeks until the babies emerge. Snake eggs being a universally appreciated delicacy in nature, the mother is doubtless subjected to many threats during her vigil. If she is to have any luck at defending herself and her children, she must have the means to detect a menace and appraise its dimensions before the interloper appears. By sensing the body heat of the approaching threat, the snake can decide whether resistance is futile and she had better slither away, or whether the moment has come to shake, rattle, and strike.

IV

ADAPTING

21

THE PLAY'S
THE THING

ALONG WITH LOVE and a good joke, playfulness seems to be something that needs no explanation, a brilliant splash of animated joy so sheerly pleasurable to watch and engage in that it carries its own justification. The whole wide thumping world is a playground, and naturalists have described their unabashed glee at the sight of a whale calf rolling and somersaulting around its mother's fluke with the viscous, goofy movements of an aquatic elephant; or a young brown bear plucking a flower with its teeth and scampering off across a meadow like a flirtatious Spanish dancer.

Yet from a biological standpoint, the problem of how play evolved and why many young mammals, birds, and even a few fish and reptiles clearly love to have fun is neither easy nor self-evident. Nothing in nature arises because it is fun, although much that we need to do ends up feeling so good that we're inspired to keep on doing it. And though play looks like a gift, a big cheery checkmark on the benefits side of life's accounting table, in fact it exacts considerable costs. The most boisterous young animals of nature, among them pronghorn fawns, Norway rats, and human children, spend on play 20 percent of the calories not needed simply to keep them alive. That's an enormous chunk of energy to invest in random and purposeless activity, the formal if not exactly mirthful definition of play. The fuel spent playing is fuel not devoted to growing up faster and getting down to the ultimate assignment of all organisms, reproduction.

Of equal significance, animals absorbed in what look like displays of youthful élan — leaping, jousting, pouncing, nipping necks, chasing phantom feathers — are taking substantial risks to their health and safety. While playing, young creatures expose themselves to predators; they attempt potentially dangerous maneuvers through treetops, near water, or by window ledges; and they court the friendly fire of a playmate's bared tooth or claw. All costs considered, evolution would never have permitted its newcomers to be so frisky were friskiness not critical to an animal's growth and performance.

Armed with heightened respect for the perilousness of play, scientists lately have taken a more sophisticated approach to their research, moving beyond impressionist observations to more rigorous studies of the physiology and psychology of play — what happens to the creature's body, brain, and behavior as it revels in its expanding prowess. They have gathered evidence that an animal plays most vigorously at precisely the time when its brain cells are frenetically forming synaptic connections, creating a dense array of neural links that can pass electrochemical messages from one neighborhood of the brain to the next. The neurons sprouting especially high numbers of synapses during an animal's days of frivolity are located in the brain's cerebellum, a cauliflower-shaped region in charge of coordination, balance, and muscle control. Through the intense sensory and physical stimulation that comes with play, links between cerebellar synapses are formed and reinforced, and those links in turn accelerate motor development. Other parts of the brain probably benefit from play stimulation as well, which may be why such big-brained species as primates and dolphins are outstandingly playful: in these creatures, the brain continues to mature long after birth and hence needs as much tweaking from the outside world as possible.

The vigorous movements of play also help in the maturation of muscle tissue. By sending varying types of nerve signals to the muscles, play ensures the proper growth and distribution of fast-twitch muscle fibers, which allow rapid muscle contractions, and slow-twitch fibers, needed for aerobic activity. Studies of muscle development in mice, rats, cats, even giraffes have all revealed that fiber tract growth and differentiation are greatest just when animals are at their most playful stage.

While the practical needs of a growing brain and muscles may account for the genesis of play, the behavior has since assumed additional, far more subtle purposes. Through play, animals can rehearse many of the moves they will need as adults; and in different species highly ritualized pastimes have evolved to suit very different needs. Young antelopes, lambs, and other herbivores play games of mock flight, bolting away from predators that are not there, a talent they clearly must master before they can safely strike out on their own.

Among carnivores like the great cats, wolves, and hyenas, cubs pretend to capture prey: stalking, pouncing, biting, swiping, tossing, growling. Little bats swoop after one another in theatrical arcs that are similar to the maneuvers an adult must employ to capture insects unawares. Young giant anteaters, among the most primitive mammals and not too impressively endowed with gray matter, nevertheless engage in elaborate play sequences called bluff charges. They puff out their hair like a cat, raise one front foot, and then hop menacingly to the side on the other three, roaring with all the fury of a clogged drain. They will repeat this ritual time and again at the age of about two months, presumably practicing the motions that will deter predators and keep other anteaters away from a prized anthill. Hatchling sea turtles, though limited by their cold-bloodedness to low-key gambols, manage to take turns holding up a front foot and vibrating it rapidly in a playmate's face, a gesture the males will use years later during courtship.

Many species adopt forms of play that are practice for both mating and for rearing young. One experiment examined the relationship between play and parental behavior by asking young rats to baby-sit. Scientists took a rat of about three weeks, the equivalent of a seven-year-old child, and put it in with a litter of newborn pups. At first, the juvenile rat tried to play with the little rodent pinklings, pounding on them and trying to wrestle with them just as it would with peers. It kabunked! kabunked! kabunked! on the motionless heap, to no avail. Within several days, however, the juvenile rat began softening up and acting like a parent, gently retrieving a pup where it crawled away, and even trying to nurse the creatures. The rodent's play behavior, it seems, is extremely plastic, and changes according to cues from the surroundings; in this case it shifted from rough-and-tumble rousting to a drill for parenthood.

Among social species, play assumes the added role of easing an animal's passage into group life, where overly selfish or hostile tendencies must be tamed, if not eliminated. Young rhesus macaques and squirrel monkeys, for example, from three months of age onward spend perhaps half their waking hours at play. The more an animal plays, the better its chances of becoming a well-integrated member of its troop as an adult. Through play bouts, an animal learns when to submit and when to pursue, and how to lose a fight gracefully. Primate play is also sexually segregated; males love to wrestle and females prefer games of chase over those of touch.

Monkeys that fail to play much while young may not end up as complete pariahs, but they are less sophisticated about maintaining alliances with other monkeys and are more mechanical in their overtures to potential mates. Play seems to make the difference in quality of life, between merely surviving and truly thriving.

Among some social species the adults play nearly as hard as their offspring, recementing, through ritual, bonds between surly creatures with excitable temperaments. The collared peccary, a highly aggressive creature related to the wild boar, plays frenetic games with its herd mates several times a week. Responding to an olfactory cue emitted by one or more herd members, the peccaries choose a spot hidden from predators and rubbed clean of vegetation. With the site chosen and the whistle blown, the peccaries, from the eldest to the smallest, begin mock-snapping, tumbling over each other, locking jaws, grunting, squealing, vaulting from side to side. Like the Caucus Race in *Alice in Wonderland,* the communal mayhem ceases just as abruptly as it began, and the peccaries settle down for a nap. The games instill group cohesion: peccaries love everybody in their herd and hate everybody else, and their ritualized sport helps affirm that xenophobic team spirit.

Most adult creatures are as stodgy as the rest of us and do not play a great deal with one another, but scientists have observed examples of extensive play between parents and offspring, once considered a hallmark of being human. The cubs of brown bears in Alaska play at least as much with their mothers as they do with other cubs, a favorite diversion being to roll around while locked in a bear hug. Gorilla mothers play peek-a-boo with their babies, and mama chimps make silly faces.

Essential to the evolution of play is the development of a language of play, a means of conveying to a potential playmate, Hey, it's time to play, or, should the game get very rough, Hey, I'm only kidding. Since many of the motions involved in play are similar to those an animal uses for less humorous purposes — killing, for example — the creature must make its playful intentions abundantly obvious. Those messages of benign intent may take on a stereotyped form. When a puppy wants to play, it will assume the familiar play bow, crouching forward on its front limbs and putting its hind end in the air.

Rats have their own play signal. To begin a game, one rat will suddenly dart away from another, stop short, and flip over on its back, a sign of willing vulnerability not usually seen in any other rodent venture. If the rat is given a drug to block the specific neural pathways in charge of commanding it to flip over, its talent for enlisting others to play will be severely hobbled.

The potential playmates will watch the manipulated animal dash from one side of the cage to the other, expecting at any moment to see the characteristic flip-flop. When it fails to come, they turn up their snouts in disdain, rather as star softball players might respond should the gawkiest kid in the class beg for a spot on their team.

For all its potential cruelty and snippiness, play among children surpasses in complexity of purpose the play of any other species. Through games, romps, and fantasies of all kinds, children practice many of the skills they will need as adults. As with other animals, playing probably has a strong physiological component, setting off the growth of synaptic connections between neurons and the maturation of different types of muscle tissue. And like other social animals, children at play learn the art of communalism, of giving and taking, of converting a momentary desire to wallop a peer into a more jocular round of roughhousing.

Children also share with other primates a marked tendency to divide up by sex and to engage in different activities once segregated. Young boys, no matter how pacifist their parents, usually enjoy vigorous physical contact, mock battles, wrestling, shouting, and the chance to turn any object longer than it is wide into an imaginary AK–47. Girls rarely play-fight and seem to prefer games requiring keen coordination, like hopscotch, jump rope, or tag — preferences reminiscent of the chase games that other young female primates love. And from culture to culture, girls often rehearse elements of motherhood, whether by playing with dolls or using each other as pretend infants. But how much of the distinction between playing styles stems from inborn predisposition and how much from socialization is one of those arguments which seem beyond resolution.

Humans also use play to master language. Infants begin with babble games, toddlers experiment with combinations of words, and older children invent stories and fantasies, and through things like tongue-twisters and riddles comment on the nature of language itself. Girls often display greater verbal fluency than boys, but whether that reflects a difference in brains or the fact that mothers speak to their female infants more than to their sons remains uncertain.

In yet another trait that distinguishes us from the rest of the animal kingdom, play bouts between parents and offspring in nonhuman species are usually initiated by the young animal. Among people, however, a baby is incapable of doing anything on its own for so long that parents must take the first step in play, wiggling the infant's foot, tickling the infant's belly, jiggling a set of keys, singing "Puff the Magic Dragon" and being re-, warded with the blinding brilliance of a baby's gummy smile. That need for the grown to play games with the young is a universal life jacket, proffering stimulation for the infant's body and brain and nourishment for the parent's weary soul.

22

HORMONES

AND HYENAS

BY POPULAR REPUTATION, hyenas are repulsive scavengers, their coats a mass of mange, their mouths rabidly agape, their laughlike calls hysterical in pitch. So it is a splendid surprise to discover, at a research colony of spotted hyenas sequestered in pens on the hills of Berkeley, that the beast is in fact a beauty, a creature every bit as baronial as that more celebrated carnivore, the lion. Its espresso-brown face is both tender and strong, familiar and alien, combining a bit of bear, a hint of snow leopard, even a flicker of harbor seal. Its rear legs are considerably shorter than its front, an improbable body plan that allows it to run long distances in pursuit of its prey, and its chest and neck are a mesh of dense muscles, which enable an animal no bigger than a border collie to easily crush the skull of a buffalo-sized wildebeest.

And pulverize prey the spotted hyena does. Far from feeding on leftover carrion, as lore would have it, *Crocuta crocuta* is the most ferocious of hunters, accounting for more game killed on the Serengeti Plain than any other meat eater. It is also the most efficient consumer, devouring flesh, bones, hoofs, teeth, fur — everything but the tips of the horns. In less than thirty minutes, a group of two dozen hyenas can reduce a five-hundred-pound adult zebra to a bloodstain on the ground. They eat so much bone that their feces look like chalk.

The most outstanding feature of the spotted hyena, however,

is its balance of hormones. While in the womb, male and female fetuses alike are exposed to dizzying levels of male hormones, particularly testosterone. As a result of the androgen bath, both sexes end up with masculine-looking genitals, the male bearing the standard equipment, the female sporting an enlarged clitoris that resembles a penis and fused, protuberant vaginal labia that look like a plump pair of testicles. Both sexes can and do get erections at the slightest excuse — when sizing up a stranger, when greeting a friend. But though the two sexes look equal, they are not: the female is in charge.

The hyena's endocrine system and its striking effect on female genitalia and manners have no parallel in the mammalian nation. Yet, as is often the case in science, the exceptions can brilliantly illuminate the rules. Biologists who study the hyena on the scratchy brown hills of Berkeley or the scrubby savannahs of sub-Saharan Africa argue that their findings will solve many puzzles of physiology and behavior, among them how androgens and estrogens jointly influence sexual development in all mammals. Already their discoveries are toppling traditional ideas about testosterone and its supposedly central role in shaping dominant behavior, suggesting that for many animals, including us, elements other than the familiar steroid hormone may be more important in the genesis and fine-tuning of a ferocious personality.

Hyena cubs, when they emerge from their testosterone-laced uterus, are the most belligerent newborns among mammals, so wired for a fight that they immediately begin attacking one another, often to the death of one. But the aggression is not only a matter of excessive male hormones. As the females age, their level of testosterone declines steeply, dipping well below that of the males, yet they remain far more pugnacious and forbidding, haughty princesses all. Indeed, when I was given a few pieces of horse meat to toss into a pen at Berkeley that held a male and a female, the male did not bother scrambling for his share of the snack — did not so much as twitch a whisker or raise a paw — until the female had had her fill.

Despite their virilized anatomy and domineering behavior, female hyenas perform their feminine roles adroitly, managing to copulate through a tiny opening in the clitoris and giving birth through that same penislike organ. Such activities look unbearably painful to observers, particularly male observers, but the events are made easier by well-timed increases in estrogen to help the skin soften and stretch.

The hyena narratives have a particular appeal for the synthesizers of biology. Most biological research breaks down into two predictable camps, each basing its ideology on matters of size: the macro versus the micro, the organismic versus the molecular. Field biology offers us the operatic complications of animals as they perform in the wilderness, on their own terms, and by their own circadian schedule. Laboratory biology serves up elegantly purified and characterized molecules — genes, proteins, hormones, the slippery lipids that sheathe the cell. The study of hyenas is one of the few attempts to straddle the divide, bringing together field observations with detailed biochemical and molecular analysis of the hormones and genes that shape animal behavior.

That convergence of attitudes may yield a reconciliation of other supposed opposites, to elucidate how the hormonal yin and yang of the estrogens and androgens operate within all of us or, on occasion, misoperate. For example, polycystic ovarian syndrome, a common disorder in which women produce abnormally high amounts of androgens and are often infertile, may be caused by a hormonal event similar to one at work in pregnant hyenas. By learning how female hyenas accommodate to high doses of testosterone without suffering ill effects, we may be led to a deeper understanding of the effects of androgens on a woman's body, a topic that remains largely unexplored.

Spotted hyenas are the largest and most common members of the hyena family, a group of four species that seem doglike but in fact are more closely related to cats and are closest to mongooses and civets. The spotted variety is the only species in which the female has virilized genitals, which is why the animal has long attracted attention, much of it disgust. The authors of a twelfth-century bestiary wrote of the spotted hyena that "its nature is that at one moment it is masculine and at another moment feminine, and hence it is a dirty brute." In his memoirs, Ernest Hemingway, an avid big-game hunter but a poor naturalist, repeated the myth that the hyenas were hermaphrodites. By the 1960s, scientists were secure in their knowledge that the female only looks masculine, but the biochemical mechanism explaining how she got that way remained to be learned.

It was in the mid-1980s that the researchers at Berkeley shipped in twenty infant hyenas from Africa to study their endocrinology and behavior. The hyenas, reared by hand in the large hillside pens, have reached their adult size, about two hundred pounds, and maintained their innate ferocity among themselves. They've taken affectionately to their surrogate parents, however, and will happily bound up on a lap for a hearty rubdown of their bristly fur. They bite only when they're caught by surprise; but they happen to be skittish enough that everybody who works with them has a few scars to prove it.

Through collated studies of the beasts in captivity and their wild comrades in Africa, biologists have learned that hyenas observe a rigid hierarchy, in which a reigning female and her offspring hold such sway that, when a clan commences tearing apart some herbivore, a strapping adult male will capitulate to the runtiest cub of the dominant female. Biologists who have followed packs of hyenas since the 1970s note that the hierarchy is dynastic: great-grandchildren of the original matriarch assume dominant positions in the tribe, and the descendents of those on the bottom are, decades later, still on the bottom. The rules of etiquette are as unchanging as the hierarchy. When two hyenas meet, they don't do so face to face, but rear to face, like a pair of shoes in a box. The subordinate member immediately lifts a leg to expose its tender genitals to the mouth of the superior hyena, the ultimate gesture of vulnerability and trust. Like an officer acknowledging the salute of a private, the dominant hyena then raises its hind leg to permit the lesser hyena to take a sniff.

With all this flashing of pendulous genitalia lying at the core of *Crocuta* social life, the question looms large: Where does the female hyena get her costume jewels? In most mammals, the male fetus takes on its masculine form courtesy of its budding testes, which release the testosterone that sculpts the rest of the genitals. The average female mammal, lacking that private fount of androgens, develops the genitals prescribed by her innate program. She further owes her feminine form to the buffering effects of the placenta. The mother has small amounts of testosterone circulating in her bloodstream — all mammals carry some amount of the opposite sex's hormone — but the placenta converts maternal testosterone into a harmless form of estrogen that cannot reach the fetus and therefore has no impact on genital growth.

The hyena, though, is not your average female nor your average matriarch. Circulating through the mother hyena's bloodstream is a large concentration of a common mammalian hormone, androstenedione, which is produced by the ovaries. Students of endocrinology have long dismissed this as a junk or inert hormone, but the hyena story suggests otherwise. The mother hyena's placenta, rather than acting as a shield against maternal hormones, takes the precursor androstenedione and transforms it into fiery doses of testosterone. The fetuses of both sexes end up exposed to levels of androgens far exceeding what even a male fetus can generate on his own.

In addition, the hyena gestation period is unusually long for a mammal of its kind, lasting about 110 days, two weeks longer than that of the much larger lion. During the extended gestation, not only do females end up with masculinized genitals, but all fetuses have a chance to mature. They grow so large that they sometimes tear the mother's clitoris as they descend through her unusual birthing organ, and they emerge with eyes open, muscles coordinated, and teeth already erupted through the gum, also unusual for a newborn mammal. The combination of exposure to testosterone and their mature weaponry is often deadly, and though hyenas are generally born in pairs, they don't stay that way for long. Most neonates root around for their mother's teat; a baby hyena roots around for the back of its sibling's neck. Within hours, one hyena usually kills the other, especially if both cubs are the same sex. Such sibling murder is quite rare among other mammals.

That infantile hostility stems almost solely from the action of testosterone. As the cubs grow, however, the hormonal and behavioral profile becomes more complex. Young males and females carry equally high doses of androgen in the blood, but the females engage in far rougher and more exuberant play than do males. And even as the females mature sexually and lose their testosterone equity, they retain their indomitable spirit. That lifelong aggressiveness suggests that hormones other than testosterone are at work in shaping the female's attitude. The most obvious candidate for such a tough-bitch compound is the precursor hormone, androstenedione, which female hyenas have in abundance and which may work on the brain in a manner very similar to that of testosterone. If this turns out to be the case, we could have a clue to the origins of aggressiveness in other female mammals. Female primates, including humans, possess significant amounts of androstenedione, which could explain why women can be highly aggressive under some circumstances, even though their testosterone levels are on average a tenth those of men.

Behind the belligerence of the female hyena is the unforgiving pace of her culture. While feeding on a fresh kill, hyenas spiral toward a frenzy, hardly stopping to take a breath between bloody mouthfuls. There is no cooperative feeding or sharing, no pass the kidneys, please. Such violent feeding behavior may well have fostered the evolution of aggressiveness among the females, who had to fight to make sure their cubs got their share. Moreover, the females form the social backbone of any hyena clan; aunts, sisters, mothers, and daughters all live together, with only a limited number of males permitted to loiter about to father their offspring. Males must disperse from the clan on reaching adolescence, and females guard their socialized territory against interloping bachelors. The structure of hyena society, then, may have favored the hormonal conditions from which rose a race of animal Amazons. But this raises yet another puzzle. Male pushiness is a problem for females of many species, so why didn't more of them take a cue from hyenas and learn how to push back harder?

23

THE WORLD'S MOST
ENDANGERED PRIMATE

THE AYE-AYE is a creature that can be described only by comparing it piecemeal with other things. It is the size of a cat, has the ears of a bat, the snout of a rat, a tail like a witch's broom, and a long knobby middle finger that would look just fine on that witch's hand. Its teeth are as tough as a beaver's and its eyes bulge out from its skull like a tree frog's. And when a baby aye-aye cries, it sounds like a squeezed rubber duck — particularly when it is being held with exceptional clumsiness by a visitor attempting neither to hurt the little beast nor to be hurt herself.

This was, after all, the first aye-aye to be born in captivity outside Madagascar, its native home, and its birth signaled a possible turnaround in the fate of the animal considered the world's most endangered primate. It was an honor for me to be allowed to pick up the three-week-old baby and feel its coarse fur, its writhing, protesting muscles, its tiny heart thumping in fear and fury.

The researchers at the Duke University Primate Center spoke graphically of the baby's efforts to bite its handlers, and of how an aye-aye's teeth can pop the top off a coconut in moments. As the twelve-ounce creature eeped shrilly and squirmed its head this way and that, I wondered whether the little darling would be injured too badly were it dropped on the floor. And when

Elwyn L. Simons, a primatologist who runs the Duke Center, decided it was time to return the baby to its mother, I nodded and enthusiastically shoved the creature into his hands as though it were a baby with diapers to be changed.

Simons is a rotund man who speaks slowly, pads about as quietly as a panther, and can imitate with delightful precision the gestures of the endangered primates he cares for. "Watch this, watch this! This is how an aye-aye drinks!" he'll cry, and then jerk his finger back and forth, back and forth, from an imaginary coconut up to his mouth, slurping slightly for effect. On biting into a banana, he grins slyly, fully aware of the portrait he presents as he eats.

His mild antics in no way detract from his solemn nature and seriousness of purpose. Simons and other scientists at zoos and universities are struggling to save the aye-aye and related lemurs from extinction. The thirty species of lemurs alive today are limited almost exclusively to Madagascar, an island off the coast of east Africa that is one and a half times the size of California. Because its forests are dwindling to the vanishing point as an impoverished and rapidly swelling human population slashes and burns the trees simply to survive, all thirty species are considered endangered. The frantic efforts to pluck the lemurs back from the precipice include breeding them in captivity, with the hope of eventually introducing at least some species back into national reserves on Madagascar; encouraging the growth of eco-tourism on the island, an approach that has worked surprisingly well in some parts of Africa and Latin America; and learning everything possible about lemur desires, habits, courtship rituals, appetites, and anything else that can be used to better the animals' chances for survival in the wild. As a group, lemurs have been much less intensively studied than, for example, chimpanzees, gorillas, and baboons.

Although lemurs are prosimians, or pre-monkeys, and in brain size and social life are considered more primitive than the so-called higher primates, they are bewitchingly vivid, some with faces like quizzical little monks, others with the shocking blue eyes of Paul Newman. The golden-crowned sifaka, a slender acrobat with a fuzzy cap of strawberry blond fur, leaps from branch to branch in strange, syrupy sideways arcs. The mouse lemur, a six-inch animal that looks like its name, is the smallest primate in the world; an extinct lemur species, the *Megaladapsis,* at six feet tall was one of the largest.

The Duke center, which has been happily successful at rearing lemurs, has more than four hundred representatives of fifteen species swinging through sixty-five acres of open-air enclosures in the North Carolina woods. There they can squabble, forage, mate, rear their young, groom each other with their comblike teeth, and otherwise live by lemur laws in conditions that roughly approximate those of the Madagascan wilderness — although the outdoor enclosures are fenced with wires that will deliver a very mild shock should a lemur develop undue wanderlust.

Of particular interest to primatologists, the lemur is a kind of living fossil, a creature that survived through the lucky accident of its geographic isolation. Elsewhere, lemurs became extinct, displaced by the bigger and more aggressive monkeys and apes; but the prosimians that migrated from the African mainland to Madagascar about fifty million years ago, floating across on vegetation, flourished without the pressure from higher primates or indeed from any significant predators. As ambulatory, sentient fossils, the prosimians hold clues to the early evolution of social behavior among our primate ancestors.

The lemurs also number among the handful of mammals in which the standard sex roles are reversed. In most higher primates, males are larger than females and frequently dominate them. Male and female lemurs, however, are similarly sized, and the female has the upper hand in their encounters, eliciting from the male displays of submissive behavior and shooing him away whenever she grows annoyed.

Whichever sex is nominally in charge, no lemur at this point is in charge of its destiny, but must rely on the conscience of strangers. Of all efforts now under way to keep lemurs alive, none is as difficult as the campaign to save the aye-aye. Not only is the primate suffering from a continuing loss of habitat, as are all the lemurs, but it has a serious image problem that makes it especially vulnerable. For most lemurs, the people of Madagascar, the Malagasy, have respect and even affection, calling them "the little men of the forest." No such pleasantries for the aye-aye. It is considered an evil omen, a harbinger of death. According to one legend, should an aye-aye point its elongated middle finger at you, you are destined to die, swiftly and horribly. To rid themselves of the curse, many Malagasy kill any aye-aye they see and place its corpse on a stake at a crossroads, in the hope that a stranger will pass by and absorb the aye-aye's malevolence.

The taboos surrounding the aye-aye are so pervasive that some think the primate, whose scientific name is *Daubentonia madagascariensis,* gained its common name as a spinoff of the Malagasy expression for "I don't know," suggesting that even to mention the creature is forbidden. One reason the animal is hated may be its outlandish appearance. The aye-aye doesn't look like any other primate on the planet; in fact, it was first classified by French researchers in the eighteenth century as a squirrel. Its long fur is a dusky, forbidding shade of black, and in the dark its yellow eyes gleam demonically. It runs with jerky, aggressive movements and looks as though at any moment it might leap at your face. It also has the dangerous habit of being curious about humans, making it an easy target for those who want to kill it. Perhaps one reason the aye-aye has not been exterminated altogether is that it is nocturnal. Most Malagasy villages lack electricity, so the people generally retire to their homes after sunset, shortly before the aye-aye begins foraging.

But the primate has its appeal. Its brain is larger and more deeply convoluted than that of any other prosimian, which suggests a somewhat greater intelligence. Its hearing is so keen that it can tap on a tree trunk and detect hollow regions within, which indicate the possible presence of the beetle grubs it covets.

The animal will then rip through the trunk with four chisel-shaped front teeth that, unlike those of other primates, will grow throughout life. And of course there is the aye-aye's extraordinary middle finger, a long thin digit that can bend in every direction, even backward to touch its forearm. The finger is an all-purpose tool for tapping the tree trunks, poking holes in eggs, pumping the liquid out of those eggs, and extracting milk from coconuts.

Since aye-ayes are night creatures, and Madagascar's long, sodden rainy season discourages most researchers, the animals have scarcely been studied, but what has been learned of late whets the appetite for more. Long thought to be solitary animals, aye-ayes in fact are quite social, building huge sleeping nests in the forks of trees and happily trading bunk beds with each other from one night to the next. They take the Kamasutra approach to lovemaking. When a female decides she is ready for mating, she will hang upside down from a branch. A male will approach her and get into position by entwining his legs around her ankles, hanging downward himself, and grasping her around the waist, letting the weight of both be borne by the female. The pair will then have intercourse for an hour or two, much longer than the usual primate session. During this dangling copulation, other males will climb up the tree and try to shove the fornicating male off so that they can get in their contributions. The female may eventually mate with more than one partner before her estrus is through. At 140 days, the aye-aye gestation period is one of the longest among lemurs.

The newborn lemur that I cuddled so stylishly at Duke had been conceived during just such gymnastics in the forests of Madagascar six months earlier, when its mother was captured and delivered to North Carolina, where it joined three others of its kind. Eventually, if the aye-ayes continue to reproduce well in captivity, the primate center may distribute several to zoos around the United States. The only zoo in the Western Hemisphere with aye-ayes is in Paris.

It will be more difficult to devise a long-term strategy for the primates on Madagascar. Since the first Indonesian settlers arrived on the island fifteen hundred years ago, about 85 percent of its spectacular tropical forest has been slashed and burned by humans for wood, farms, and grazing land. The island's thin soil is badly eroded and its nutrients depleted, further threatening the forests that remain; and the human population continues to grow at 2.1 percent a year, one of the fastest rates in the world.

Enormous international will now exists to rescue Madagascar, which is lush with thousands of life forms found nowhere else, including 142 species of frogs, 106 types of birds, 6000 species of flowering plants, and half the world's chameleons. But whether the wealth of species can be preserved while so many Malagasy remain impoverished, and whether the people can learn to view the aye-aye as lovable rather than unmentionable, remain frighteningly open questions.

24

PLENTY OF
FISH IN THE SEA

THE DATE IS A DUD, and both parties know it. Yet as long as they are stuck with each other for a time, they make a wan effort to flirt. He lunges lazily toward her. She quivers gently in response. He flaps his tail against her. She flares her gills to show their provocative red undersides. He circles around, charges her again, and tries to nip her, but now she's bored with the charade and moves away from him. Reacting likewise, he drifts off to the opposite end of the tank. For a few moments, each is lost in the inscrutable vastness of fish thought. And then it happens. The female opens her full, sensuously carved lips into the widest, roundest, most perfect, least courteous gape of mouth that can be imagined: a fish yawn.

"She doesn't seem very interested, does she?" Suzanne Henson, a graduate student, says to me, with only a trace of drollness. Her pen is poised in front of her notebook to take notes on the fishes' behavior. The pen does not move; she sees nothing worth noting.

This is not as it could be. The fish before us are Midas cichlids, known for their vigorous, brutal mating dances, which can verge on S & M pornography. When a female is excited, she does the "slip motion," gliding her entire body along the body of the male. Her genitals swell and she grows heavy with eggs. For his part, the libidinous male is an abusive male. He'll smack the female with his tail. He'll bite her so hard you can hear the crunch. He'll lunge again and bite again; she'll slip-glide again and flash her rosy gills again.

That, at least, is cichlid love at its height. But not now, and not with these two slugs. Their disastrous date is finished, the experiment over, and each is returned to its proper tank.

The study I witnessed is part of an effort at the University of California in Berkeley to understand the mating choices and customs among the Midas cichlid (pronounced SICK-lid), a beefy, square-jawed fish from Nicaragua that comes in two color schemes, zebra-striped or gold — the latter accounting for the species name. Like many other cichlids, the Midas are the marrying kind, forming pairs that last as long as the fish do, and the Berkeley scientists are seeking to understand what inspires a Midas fish to choose one partner over another.

The question is part of a broader consideration of the sexual, social, and feeding behaviors of cichlids, a wildly diverse family of fish whose traits could lead to an understanding of the abiding mysteries of how species evolve and how variety arises in nature from monotony.

More than a thousand species of cichlid live in the lakes and rivers of Africa, Madagascar, India, and Latin America. They are a highly successful tribe, frequently dominating their environment through a blend of intelligence — said to be unusually high for a fish — and elaborate rituals of parental care. But what makes them so unusual is the number of species that can coexist in the same place. Over five hundred varieties of cichlids swim in Lake Malawi, in southeast Africa, and about two hundred other species live in Lake Tanganyika, in Tanzania. Some species are bigger than goats; others could fit in a thimble. Some are thick and boxy; others lean and long. They are brown or turquoise or every shade of a neon rainbow painted on a single beast.

The cichlid's rate of speciation has been explosive. In Lake Victoria of East Africa, for example, three hundred species arose from one progenitor species in less than 200,000 years, an evolutionary pace that no other animal group has rivaled. Certainly none of the other fish groups found in the three African lakes has undergone anything approaching the spectacular diversification managed by the cichlid family.

Scientists have long been captivated by these fish, seeing in them a far greater opportunity to probe essential evolutionary patterns than was afforded by another famously diverse family, Darwin's finches. Much of the research has relied on traditional taxonomic and observational approaches, the tallying up of species through studies of fish anatomy and fish behavior. More recently, biologists have added molecular analysis to their research, tracing cichlid lineages and cichlid radiations by studying the fish's DNA. The genetic approach has confirmed previous results from taxonomy: cichlids are monophyletic, that is, they all originate from a single ancestral fish that arose perhaps 120 million years ago, when India, Africa, and Latin America formed one giant landmass. Since the breakup of the continents, the founder fish that were carried off to different regions of the earth have gone their separate ways, speciating rapidly in all cases, yet by distinctive genetic mechanisms, from one lake or river to another.

In some instances, species that look and behave radically differently from one another turn out to be almost identical genetically. One genetic study looked at the DNA of fourteen Lake Victoria cichlid species exhibiting highly divergent feeding behaviors: a snail eater, a cichlid that feeds on fellow cichlids, a cichlid that eats only the eyes of other cichlids, another that exclusively sucks young cichlid fry out of the protective mouths of their parents. Yet despite the fishes' specialized appetites, their genes differ by only two or three base pairs, or chemical subunits, out of the many thousands that constitute the genes examined. There is more genetic variation among people than there are among these fourteen fish species — and people, keep in mind, are all members of the same species.

Such findings suggest that much of the success of the cichlid family lies in its unusual degree of molecular flexibility, with minor differences in genes yielding enormous disparities of comportment. And it is the cichlid's ability to specialize that helps explain how so many species can live cheek by gill in the same body of water, with each still managing to earn a living. If all cichlids were bottom grazers one species would likely outcompete the others into oblivion. But each cichlid has evolved its own hunting method, and every strategy strains credulity. One cichlid resembles a rotting fish and spends a lot of time floating as though dead; but when another fish approaches, thinking it has happened on an easy meal, the corpse springs to life and attacks the would-be scavenger.

A cichlid in Lake Tanganyika has its head bent permanently to the left, an adaptation that enables it to scrape, with its teeth, a meal of scales off the right side of a passing fish's body. Another species has evolved a head curving to the right, the better to shave scales from a prey fish's portside. The cichlid overturns the standard idea in ecology that there are various niches waiting to be filled; instead, it takes the approach of a good entrepreneur and creates its own niche from scratch.

Most scientists believe that the original cichlids in the great African lakes were generalists that became specialists as competitive pressure increased. The question of how they were able to speciate so swiftly and broadly remains a great conundrum among ichthyologists, but the cichlids undoubtedly were aided by their unusual jaw configuration. They have one jaw set in their mouth, as the average fish does, and a second in their throat. With the throat jaw available to process food, the mouth jaw is freed of certain basic physiological constraints and can evolve very specific methods for capturing prey. In essence, the throat jaw is a jack-of-all-trades, the mouth jaw a master of one.

Variations in dining strategies, however, are not the family's only distinguishing traits. Fish hobbyists love cichlids, in the main because they admire the fish's famed courtship and fry-rearing practices. Most fish lay eggs and leave them, or the father may remain to watch the eggs until they hatch. Among many cichlids, both parents engage in protracted parental care. They brood the eggs in their mouths, and even after the fry are born, they protect the little fish by taking them back into their mouths when predators approach, sucking the fry in as though sucking in strands of spaghetti.

This habit of mouth brooding has given rise to a few outstanding features on male cichlids. Because predatory pressure in a cichlid's habitat can be relentless, many females, after laying their eggs, frantically turn around and begin scooping them into their mouths — before the male has had a chance to fertilize anything. Males have adapted to this by evolving bright spots on their rear fins that strongly resemble eggs. When the female has finished sucking in her clutch, the male gives his rear fin a shake. Thinking she has left a few ova behind, the female tries nipping at the decorated appendage, at which point — *whoosh!* — the male releases a stream of semen into the female's open mouth. In some species, both parents feed the fry with their own flesh, allowing the young fish to nibble at the scales and nutritious mucus cells beneath. The fish, in other words, become a couple of floating breasts.

In light of the high investment the parents make in their offspring, they have much incentive to select the worthiest possible partners. Here is where the experiments with the Midas cichlid offer a few clues. It happens that the two subtypes of the species are not fixed. About 8 percent of the time, a striped Midas cichlid changes its coat as it grows and becomes bright gold. When given a choice, everybody — zebra-striped and gold alike — will choose a gold partner over the more common striped variety. That could be because the golds look more threatening. Cichlids must often fight off outsiders while rearing their brood, so toughness in a mate is highly valued. Through their detailed matchmaking trials at Berkeley, the biologists have learned that mate choice proceeds in two steps. First, a female finds a male who appeals to her, either because he has the right color, the right fishly odor, or other trait that has yet to be determined. But once she has demonstrated her liking for him, he exerts his own choosiness by behaving most aggressively toward her. If she has any hope of winning his affections, she must threaten him back with equal ferocity. A female who is intimidated by a male is a female destined for failure. However, once the male has determined that the female is tough enough, he will mate with her and treat her gently ever after.

The odds of a male and female cichlid sharing just the right color and chemistry are slim, which is why many a Midas encounter ends in ennui and a giant fish yawn.

25

CHASING CHEETAHS

THE CHEETAH may be a gorgeous Maserati among mammals, able to sprint at speeds approaching seventy miles an hour, yet it has not been able to run away from its many miseries. Once, the cat ranged throughout the African continent, the Near East, and into southern India; now, it is extinct almost everywhere but in scattered patches of the sub-Sahara. Farmers and ranchers in Namibia shoot them as vermin. On reserves, where cheetahs are often forced into unnatural proximity with other predators, they are at the bottom of the meat eaters' grim hierarchy; lions will go out of their way to destroy cheetah cubs, while hyenas, leopards, and even vultures can easily chase away a cheetah from its hard-caught prey. To make the magnificent cat's story more poignant still, many scientists have concluded that the species is severely inbred, the result of a disastrous population crash thousands of years ago from which the poor brute has hardly had a chance to recover.

Studies of cheetah chromosomes have shown a surprising lack of genetic diversity from one individual to the next, and as a result the cheetah has been widely portrayed as sitting under an evolutionary guillotine, the population so monochromatic that, in theory, a powerful epidemic could destroy many if not all of the fifteen thousand or so cheetahs that survive in the wilderness.

Some zoos have complained that their cheetahs are infertile, and they have attributed the problem to the cheetah's bleak genetic makeup, calling into question the long-term prognosis even for cats living in the pampered confines of a park. Now some maverick biologists argue that this widely held notion of the inbred cheetah may be wrong, an artifact of test-tube manipulations with little relevance to the cat's workaday world. They insist that, far from displaying the negative effects of inbreeding seen in other animals known to be genetically homogeneous, like certain strains of laboratory mice or pedigreed dogs, cheetahs are in many ways robust, more like ordinary house cats than the feeble product of generations of incestuous couplings.

The significance of the debate extends far beyond the spotted Concorde of a cat. Scientists are seeking to calculate the odds that any number of endangered or threatened species are likely to survive into the twenty-first century, and among the many questions they ask is how much genetic diversity a creature requires if it is to rebound from the brim of extinction. Inbreeding is thought to be harmful to a species for two reasons: first, it allows hazardous recessive traits to come to the fore, resulting in birth defects, stillbirths, and in some cases infertility; and second, it leads to a genetically uniform population without the diversity to resist epidemics and environmental changes. When strains of laboratory mice are inbred repeatedly to exaggerate a trait under study — for example, a propensity for breast cancer — the rodents end up comparatively lethargic and dull-witted, prone to miscarriages and the birth of mutants.

But the scientists who dismiss the inbred cheetah dogma say their zoo cats almost never bear defective cubs, are perfectly fertile and vigorous, and have great variation in their immune systems. While the cheetah may look genetically tenuous when its DNA is appraised, by such real-life measurements as fecundity, litter size, cub health, and immune response, the cheetah is perfectly fit for the next millennium. The work calls into question the validity of a strictly molecular approach to the sometimes murky science of species preservation, and it strongly suggests that scientists do not yet understand why certain genetic patterns detected in laboratory tests translate into the genuine strengths and weaknesses of a wild animal. The work also indicates that zoos having trouble propagating cheetahs in captivity perhaps should not blame the animal's DNA, but rather their own ineptitude at animal husbandry and matchmaking. Indeed, at some zoos cheetah breeders have been too successful in their ministrations, resulting in a population boom and subsequent demands for a feline version of Norplant.

Persuasive though the arguments of the neo-cheetah champions may be, their evidence is hardly airtight. Cheetahs do display a remarkable lack of genetic diversity compared with, say, tigers or leopards. When skin from one cheetah is grafted onto another, it takes an extraordinarily long time for the immune system of the transplant recipient to reject the added flesh — strong evidence that cheetahs are practically clones of one another. And though cheetahs may begin life with the strength to outrun all contenders, they end life rather too quickly for a large mammal. Even in zoos, cheetahs rarely live beyond seven years, a third of the life span of other exotic cats in captivity. Of course, it's possible that the cheetah is a naturally short-lived species, but it's equally possible that the cat suffers chronic and ultimately fatal health problems as a result of overall genetic frailty. The commonest cause of death among cheetahs in captivity is kidney failure, a condition that may or may not be associated with rotten DNA.

Just as the consequences of the cheetah's genetic homogeneity are debated, so too are the causes. By one scenario, the cheetahs were some of the earliest victims of humanity's depredatory approach to environmental management. At the end of the last ice age, ten thousand years ago, humans advancing rapidly in the wake of the retreating glaciers supposedly wiped out cheetahs everywhere but in a few pockets of Africa. In that mass extermination, the scenario goes, cheetahs lost more than 90 percent of their genetic variation, suffering a catastrophic population bottleneck from which it has only barely begun to recover.

In another and equally plausible explanation, the cat may lack genetic diversity not because of the cruel hand of the human hunter, but because it is the most specialized cat of all, with a body designed from snout to spine for the sole purpose of running at supermammalian speeds. By this argument, the evolutionary process that focused on enhancing the cat's capacity to sprint ended up throwing out a lot of other genes along the way. In other words, the business of being a cheetah could require genetic homogeneity, and a modest life span could be part of the package deal.

The cheetah is a spectacular example of streamlined design. It is relatively petite and light-boned, weighing only about seventy pounds; it has an aerodynamically small head, unusually long legs, a flexible spinal column, and a sliding shoulder to lengthen the stride. Its canine teeth are very small, leaving plenty of room for its nasal passages, which are wide enough for the animal to take in a lot of oxygen. The cheetah hunts not by stalking prey, but by bolting at its quarry in an explosion of energy so exhausting that the cat has to wait fifteen to twenty minutes, panting, before it can eat.

Because the cheetah is slighter than most African carnivores and lacks large canines to defend itself, it cannot ward off competing meat eaters that want its dinner, so when confronted, it will usually give up and skulk away. In fact, the cat is so unaggressive by nature that when I entered a large pen at the San Diego Zoo to see a mother and her five young cubs, the cats allowed me to approach to almost within stroking distance, the mother looking on with a mixture of boredom and irritation, the cubs cutely raising their fur and hissing ever so slightly. Had the cage held a family of tigers, I might not be around to celebrate cheetahs.

In the final analysis, the cheetah's long-term future very likely rests not on genetic research, but on old-fashioned remedies like preserving its remaining habitat and enlisting the help of those who live alongside it. In Namibia, where the cheetah does not have to compete with many other carnivores, as it does elsewhere in Africa, the feline fares reasonably well, and its biggest problem is ranchers who shoot it in the belief that the cat threatens their livestock. Biologists in Namibia are seeking to convince the cattle owners that cheetahs in fact kill very few livestock animals, and to establish a compensatory program should a calf occasionally be lost. With its last sizable free-ranging population now confined to Namibia, the cheetah is being pitched as a unique Namibian cat and thus a source of national pride. More than an ideal genetic profile, the cheetah needs a bit of panting room and all the P.R. its noble bearing can buy.

26

BUSY AS A BEE?

IN THE LANGUID DAYS of midsummer, or during the bittersweet slump between Christmas and New Year's, or on any afternoon that beckons with beauty, those of you who feel the urge to take it easy but remain burdened by a recalcitrant work ethic might do well to consider that laziness is perfectly natural, perfectly sensible, and is shared by nearly every other species on the planet.

Giving the lie to the old fables about the unflagging industriousness of ants, bees, beavers, and the like, a new specialty known as time budget analysis reveals that the great majority of creatures spend most of their time doing nothing much at all. They eat when they must or can. They court and breed when driven by seasonal impulses. Some species build a makeshift shelter now and again; others fulfill the occasional social obligation, like picking out fleas from a fellow creature's fur.

But more often than not, animals across the phylogenetic spectrum will thumb a proboscis at biblical injunctions to labor and proceed to engage in any number of inactive activities: sitting, sprawling, dozing, rocking back and forth, ambling around in desultory circles. If you were to follow an organism in the field for an extended period of time and catalogue its every activity for every moment of the day, you would probably reach the conclusion that, by George, this thing isn't doing much, is it? In fact, compared with other creatures, human beings spend anywhere from two to four times as many hours working, even more when family and household duties are taken into account.

Lest we feel smug about our diligence, however, a fair analysis of animal inactivity shows it is almost never born of aimless indolence, but instead serves a broad variety of purposes. Some animals lounge around to conserve precious calories, others to improve digestion of the calories they have consumed. Some do it to stay cool, others to keep warm. The hunted is best camouflaged when it's not fidgeting or fussing, and so too is the hunter, who wishes to remain concealed until the optimal moment for attack. Some creatures linger quietly in their territory to guard it, and others stay home to avoid being cannibalized by their neighbors. So, while there may not be a specific gene for laziness, there is always a good excuse.

The possible reasons for laziness are so diverse that some biologists are shifting the focus of their research. Rather than observing the behavior of animals in action, as field researchers historically have done, they are attempting to understand the many factors that lie behind animal inertia. They hope that, by learning when and why an animal chooses inactivity, they can better understand such central mysteries of ecology as the distribution of different species in a particular environment and how animals survive harsh settings and lean times. As one time budget analyst put it, "I used to focus on movement, foraging, mating behavior. Now I wonder about why animals sit still."

Animals certainly give their researchers much to ponder. Those who have studied lions in the Serengeti for the past twenty years say they spend nearly all their time staring through binoculars at tawny heaps of fur; the pride's collective unconsciousness is broken only by the intermittent twitch of an ear. Lions can stay in the same spot without budging for twelve hours at a stretch. They're active on their feet for perhaps two or three hours a day. In that brief spate of effort, they are likely to be hunting or devouring the booty of that hunt, which is one reason they need so much down time. A lion can eat an enormous amount in one sitting, maybe seventy pounds of meat. Its stomach bloats so hugely that, by the time it is through, the mighty king of the jungle can barely stagger over to the shade of a tree, where it collapses, belly up, for a catnap that might more properly be described as a coma.

Monkeys are commonly thought of as nature's indefatigable acrobats, but many species sit around as much as three quarters of the day, not to mention the twelve hours of the night they usually spend sleeping. Primatologists studying the woolly spider monkey in Brazil were amused to discover its lax habits. One morning they awoke before dawn to get out to a distant observation site by seven A.M., when, they assumed, the monkeys would start their day's foraging. The scientists arrived on schedule, set up their equipment, and waited for the excitement to begin. They waited and waited, doodled in their notebooks, wondered idly whether the monkeys had chewed a few too many coca leaves the night before. By eleven o'clock, the monkeys were still sleeping, at which point the researchers nodded off themselves.

Hummingbirds are the world's most vigorous and energy-intensive fliers — when they're flying. The birds turn out to spend 80 percent of their day perched motionless on a twig; at night, they sleep.

Beavers are thought to bustle about so singlemindedly that their name has been verbified into a synonym for work. But beavers emerge from the safe haven of their lodge to gather food or to patch up their dam for only five hours a day, give or take a few intermissions; even when they're supposed to be most active, they'll retreat into the lodge for periods of rest. The spade-foot toad of the Southwestern desert burrows three feet underground and refuses to budge for eleven months of the year. In that time it does not eat, drink, or excrete waste, all the while conserving energy by turning down its core metabolism to one fifth of what it is during its single active month. If you were to happen on one of these dormant amphibians while digging in your cactus garden, you could scoop it up as easily as you would a rock or a potato.

Even the busy bees or worker ants of Aesopian fame dedicate only about 20 percent of the day to doing chores like gathering nectar or tidying up the nest. Otherwise, the insects stay still, as though they'd misplaced their to-do list and frankly didn't give a damn. The myth of the tireless social insect probably arose from observations of entire hives or anthills, which are little galaxies of ceaseless activity. But now that scientists have learned to tag individual insects to see what each does from one moment to the next, they find that any single bee or ant has a lot of surplus time.

Biologists studying animals at rest turn to sophisticated mathematical models, resembling those used by economists, which take into account an animal's energy demands, fertility rate, the relative abundance and location of food and water, weather conditions, and other factors. They do extensive cost-benefit analyses, asking questions like: How high is the cost of foraging compared with the potential calories that may be gained? Such a calculation involves not only a measure of how much energy an animal burns as it rummages about relative to what it would spend resting, but also a consideration of, for example, how hot it will become in motion, and thus how much of its stored water will be needed to evaporate away heat to cool the body. Once they complete their computations, the biologists usually acknowledge their respect for the animal's decision to lie low.

Some strenuously object to the word *laziness* to describe any animal behavior; they say it implies the willful shirking of a task that would improve the animal's lot in life if it were done. Animals are inactive when they have to be. The moose, a ruminant, has to stay fairly still so that its four-chambered stomach can metabolize its fibrous diet of leaves, stalks, and grasses. For every hour of grazing, the moose needs four hours for digestion. And it has other good reasons to avoid overzealous activity. A moose is huge and works up quite a sweat as it snuffles about in the bushes. Were it to forage past a reasonable point and raise its body temperature close to the lethal maximum, it would be in serious danger should a predator appear. The act of running away would push the moose's core body temperature up over the top, leading to death by heat stroke — and an unearned windfall for its pursuer.

Researchers who have looked at hummingbird behavior conclude that the tiny birds are also perfectly justified in taking frequent breaks. To hover in midair while sipping from long-tubed flowers, they must beat their wings in elaborate figure-eight patterns at a rate of sixty times a second. The costliness of their flight would give pause even to NASA; the action burns more fuel in calories per gram of body weight than any other bit of animal athleticism ever studied. Flying is so draining that many hummingbirds and their African counterparts, the sunbirds, are better off staying motionless unless the food they can obtain is very rich indeed. To help arrange for dinner without having to travel too far, sunbirds will choose a territory and stand around the perimeter, waiting for the flowers within to become plump with nectar.

For some creatures, immobility carries so many benefits that they become almost Buddha-like in their stillness. The fringe-toed lizard, which lives in the desert of the southwest United States, sits motionless just below the surface of the sand for hours, with nothing sticking up but its eyes. As the lizard sits, the sand warms and invigorates it, lending it the necessary charge to lurch out at any edible insect that passes by. Should it instead see a predatory snake approaching, the lizard can initiate its own near-death experience, temporarily suppressing its breathing and stopping its heartbeat. Finally, by staying snug in its sandy blanket, the lizard cuts down on water loss, a constant threat to desert creatures.

In a harsh place like the desert, most animals spend much of the time waiting for water and coolness. That spade-foot toad of the Southwest comes out only in July, when the annual rains bring insects to feed on. Male and female toads meet and mate the very first night they emerge from their rocklike state, and they begin eating enough to put on the extra 30 percent in body fat required to make it through their dormant eleven months.

Several hundred species of mammal go into hibernation each winter, cutting down on energy expenditure by dramatically lowering their metabolic rates. When a ground squirrel hibernates, its heart rate slows to only one or two beats a minute, and its body temperature descends to near freezing. For herbivores, winter hibernation makes sense: there's nothing to eat, the weather's bad, reproduction is inadvisable, and there are still predators about. The best thing to do is go into suspended animation.

Sometimes a biologist is stumped by apparent indolence that cannot be explained by obvious things like inclement weather. Students of the naked mole rat, a homely, hairless, blind little social mammal that spends its entire life underground, long wondered why the largest mole rats in a group did the least and seemed to sleep the most. They found out one day when they introduced a snake into the colony set up in their lab. The big mole rats immediately sprang up and attacked the snake en masse. Though apparently asleep, they were just maintaining quiet vigilance.

Such a need for vigilance may explain why bees and ants spend so much time resting. Honeybees have a so-called soldier caste of workers, a standing army that does little or nothing around the hive but is the first to act if the hive is disturbed. Other bees and ants may be saving their energy for a big job, like discovering an abundant new source of food, which will require overtime effort to harvest, or the intermittent splitting of one hive into two, which will suddenly leave fewer workers to do the same tasks.

New studies show that social insects cannot afford to waste their energy on trivial activities. Ants and bees are like nonrechargable batteries, born with a set amount of energy to devote to their colony. They can use up that energy quickly, or they can use it slowly, but they can't get more of it by eating right or exercising regularly. In other words, the harder they work, the sooner they die. With that thought in mind, one can sympathize with a bee's desire to take a moment to stop and not sniff the flowers.

And perhaps a bit of luster can be lent to the much-maligned creature that gives laziness its synonym: the sloth. Found throughout Central and South America, the sloth hangs from trees by its long, rubbery limbs, sleeping fifteen hours a day and moving so infrequently that two species of algae grow on its coat and between its claws. A newborn sloth sits atop its mother's belly and is so loath to move that it freely defecates and urinates on her fur, which she will only intermittently bother to clean. But lest you find such sluggishness perverse, note that the sloth is suited to its niche. By moving so slowly, it stays remarkably inconspicuous to predators. Even its fungal coat serves as camouflage. With the algae glinting greenish-blue in the sunlight, the sloth resembles the hanging plant it has very nearly become.

Humans generally spend more time working than do other creatures, but there is considerable variability in industriousness from one human culture to the next. The average French worker toils for 1646 hours a year, the average American for 1957 hours, and the average Japanese for 2088.

One reason for human diligence is that we, unlike other animals, can often override our impulses to slow down. We can drink coffee when we may prefer a nap, or flick on the air conditioning when the heat would otherwise demand torpor. Many humans are driven to work hard by a singular desire to gather resources far beyond what is required for survival. Squirrels collect what they need to make it through one winter; only humans worry about college bills, retirement, or replacing their old record albums with compact discs.

Much of that acquisitiveness is likely to be the result of cultural training. Most hunter-gatherer groups, who live from day to day on the resources they can kill or forage, and stash away little for the future, generally work only three to five hours daily. Indeed, an inborn temptation to slack off may lurk within even the most work-obsessed people, which could explain why sloth ranks with lust and gluttony as one of the seven deadly sins.

27

IT'S ONLY

HUMAN

FOR ALL its secular posturing, science has in common with many religions a zealous adherence to the concept of sin. There are the deadly scientific sins, like fabricating results or failing to give proper credit to one's peers; and there are the little sins, like experimental sloppiness or appearing once too often on television. Among researchers who study the behavior and ecology of nonhuman animals, the biggest little sin of all has long been the dread practice of anthropomorphism: to ascribe to the creature under scrutiny emotions, goals, consciousness, intelligence, desires, or any other characteristic viewed as exclusively human.

By the traditional dictum, a scientist should never presume that an animal has intentions or is aware of what it's doing or even feels pain. The truly objective biologist will refrain from projecting personal feelings onto the animal, and instead confine the research to a rigorous collection of observations and a dispassionate statistical analysis of the data. Lately, however, a growing contingent of animal behaviorists has broken ranks and proclaimed that anthropomorphism, when intelligently and artfully done, can accelerate our understanding of the lives and sensibilities of the beasts that surround us.

Nobody is suggesting that animals are just little people with feathers or fur who don't happen to be on the Internet. What the anthropomorphists argue is that many species give strong indications of possessing self-awareness, awareness of others, and a certain degree of foresight and intention, all traits that the orthodox antianthropomorphists insist animals do not have. The researchers say that by granting their nonhuman subjects a measure of motivation and desire, they can ask comparatively more compelling questions and devise more revealing experiments. "I'm not going to make any outrageous claims that animals are doing a lot of conscious manipulation of their surroundings," said one ardent anthropomorphizer, "but I am going to claim without hesitation that animals form expectations about the future."

The evening grosbeak, for example, congregates with its kind in a circle rather than in a line or a haphazard gaggle. By taking an anthropomorphic slant on the puzzle of grosbeak grouping behavior, biologists determined that a circular formation makes it easy for the birds to pay close attention to interflock cues. If one bird sees that another is keeping vigilance for predators, it will start eating, freed of the need to watch its backside. How else to explain the behavior but to assume that the grosbeak expects and believes that its scanning of its peers has done it some good?

The neo-anthropomorphists say there is no scientifically valid reason to suppose that an unbreachable gap separates us from the rest of the natural world. The human skeleton, the human body, the molecular workings of the human cell, the nerve tissue that makes up the exalted human brain all look remarkably like those of other species; what hubris leads us to assume that human psychology and human behavior sprang up *de novo*, bearing no resemblance to the behavior of any other being on earth? We are living organisms on a continuum with other organisms, the anthropomorphists proclaim, and it makes no sense to deny aspects of our humanness when looking at the behavior of our fellow life forms.

The New Guard also argues that anthropomorphism is really just another word for empathy, the willingness to delve into what has been called the "private experience" of the animal — the sights, smells, roars, and buzzes that define its world; the things it notices and the things it ignores; how it reacts when confronted by the unexpected. A good anthropomorphist will try to crawl inside a creature's skull to see the world as it does — to slither like a black-tailed rattlesnake, to soar like a red-shouldered hawk — and this practice demands that you believe the skull a worthy and complex place to be.

In one experiment with a distinct anthropomorphic flavor, biologists attempted to gauge the awareness of the hognose snake. The snake will go to many and varied extremes to deter predators. When initially confronted, it puffs up like a venomous cobra, though it lacks any venom. If that doesn't scare off the enemy, it goes into a seizure, writhing, contorting, rolling over, defecating, and finally turning on its back, its breathing stopped and its tongue hanging out as though it were dead. The performance was long viewed as evidence of simple fright, but biologists proved that in fact it is an act of deception: the snake is playing dead. And it plays dead to a cunningly refined degree. If a person stands near the snake and stares at it, the reptile remains motionless, upside down, tongue extruded; the moment the person's eyes are averted, the snake flips onto its belly and slips away.

The iconoclastic attitudes have sparked a lively debate among animal behaviorists and generated many rounds of point and counterpoint. John S. Kennedy, of Oxford University, threw down the gauntlet in 1992 in a slender volume called *The New Anthropomorphism,* in which he denounced the practice of anthropomorphism as a kind of genetically programmed "disease" that must be cured if the field of animal behaviorism is to survive. Stinging volleys against the book and its philosophical ilk have since followed in numerous ethology journals, as proponents of anthropomorphism criticize Kennedy for his ignorance of their work on animal intelligence, animal language, and animal awareness.

Most scientists fall somewhere between the extremes, accepting that animals do possess a more complex inner life than they have been given credit for, but fearing that the most enthusiastic anthropomorphists take too broad a brush to their topic and append too many of their own impulses and beliefs on to their animals. In a particularly vivid example of the hazards of viewing nature through excessively Homo sapienized lenses, the great elephant watcher Cynthia Moss described her first encounter with what she initially called green slime disease. Early in her observations of the behemoths, she noticed that all the male elephants in the group were getting green slime on their bodies, particularly around their penises. Finding the apparent venereal goo repulsive to behold, she assumed the female elephants would likewise be put off by their viscous mates. But as she continued her studies, she learned that the slime was in fact part of the males' rutting display, a sign of their sexual randiness and readiness that the females happened to find attractive. In a similar vein, rank anthropomorphism can lead to inappropriate animal care, such as keeping a rat's cage antiseptically clean, when in fact rodents are accustomed to wallowing in a rich diversity of odors.

The anthropomorphism dispute also brings forth a bit of species-ism among the contestants, each defending his or her animal as the cleverest and most sophisticated and hence the worthiest of being anthropomorphized. In attacking Kennedy, some anthropomorphizers revealed their anthropocentrism, as they sneeringly said that his attitude was precisely what you would expect of an entomologist who had spent his life studying lowly (i.e., inhuman) aphids. Primatologists argue that it makes perfect sense to detect similarities between the behavior of chimps and our own antics, since the two species share at least 98 percent of their DNA. But those who study animals that are more distantly related to us insist their creatures also display socially sophisticated behavior abrim with awareness and intelligence. Sheep are commonly thought to be as woolly in brain as they are in body. Yet biologists who specialize in sheep behavior have observed their animals engaging in such socially complex behavior as making conciliatory gestures after a fight and sticking up for a sheepmate when the animal is persecuted by the group.

. . .

The debate over humanity's relationship to the animal kingdom is not new, of course. The Judeo-Christian tradition gave man the role of God-proxy and dominion over the beasts. In the seventeenth century, René Descartes claimed that beasts could not think and hence barely were — and certainly had nothing in common with rational man. Nevertheless, people have always anthropomorphized the creatures around them, particularly the animals they liked. When Charles Darwin formulated his theories of natural selection and the descent of man, he found that his nineteenth-century audience objected to the idea that they were related to apes; and, in fact, the creature he described in his writings as most humanlike was the dog.

By the 1920s, members of a school of thought called behaviorism began to hold sway, arguing that there was no need to assume an animal had an inner psychological state — that all activities could be described objectively for what they were, without reference to emotions or motives. (Psychologists even attempted to apply similar mechanistic principles to the study of human behavior.) But in the 1970s, Donald Griffin, the discoverer of echolocation in bats, began to question the premise that animals lacked consciousness, and the dike of strict behaviorism started to crack. At the same time, the brazen school of sociobiology came into being, with its emphasis on evolutionary motives for a gamut of complex social behaviors. Suddenly, it became permissible to consider that animals might scheme, conspire, and maneuver, all for the sake of perpetuating their genetic legacy.

The debate now centers on the degree to which animals are aware of what they do, and the degree to which scientists should care. Some animal behaviorists who criticize the anthropomorphists argue that scientific techniques for studying brain function are far too primitive to get at the question of animal intentions. If an animal can't tell us what it intends to do, and if there is no method for visualizing that intention, what proof do we have that the intention is there? By talking about animal purposefulness as though it were in the realm of science, they complain, we're not only putting the cart before the horse, we're pretending to do it at the behest of the horse.

The critics admit that it's impossible for an ethologist not to form strong bonds with the creatures under study; but good scientists, when presenting their data for wider consumption, will clean up their subjectivity and stick with solid data and impartial analysis. Moreover, they say, to assume an animal is aware of its behavior violates a prime directive in science, that one should always seek the simplest explanation for any observation, and the simplest assumption is that most behavior doesn't require awareness or animal emotions or strategic planning. After all, it was noted half a century ago that if microbes were blown up to the size of a cat or a dog, they too would look as though they had wishes, feelings, and intelligence, so thoughtful does their behavior seem as they glide toward a drop of sugar water or away from a noxious chemical.

In response, anthropomorphizers declare that they're the ones who make the simplest, cleanest, and most sensible assumptions about animals, and that it's the anthro-chauvinists who repeatedly contort themselves into untenable positions to preserve human exclusivity. Years ago, the chauvinists argued that only humans use tools. Then it was discovered that some animals, like chimpanzees and elephants, use sticks, rocks, and other objects as tools. So the naysayers said only humans are capable of changing the tools they use; then it was found that chimpanzees alter their sticks depending on what they need to do with the devices. The latest argument is that only humans use tools to make other tools. And so the definition of our *sine qua non* bends, squirms, and adapts in the face of scientific discovery. The need to feel special is, after all, only human.

V

HEALING

28

A NEW THEORY OF

MENSTRUATION

THE MENSTRUATING WOMAN has been variously vilified, feared, pitied, or banished from the village to spend her bloody days in solitude. Even the standard medical explanation connotes loss. A woman bleeds each month, the story goes, as a way of discarding her unfertilized eggs and the uterine lining that had been optimistically fattening up in anticipation of a baby that never arrived. When the womb sits empty, the womb must weep.

Well, women, delight; you have nothing to lose but your shame. An evolutionary biologist has proposed a radical new way of viewing menstruation, one that gives an active and salutary spin to the dirty business of having a period. Margie Profet of the University of Washington, a scientific iconoclast of the old, rattle-their-bones revolutionary school, suggests that menstruation evolved for women's own good, as a mechanism for protecting the uterus and Fallopian tubes against harmful microbes delivered by incoming sperm.

According to this reading, the uterus is extremely vulnerable to bacteria and viruses that may be hitching a ride on the sperm, and menstruation is an aggressive means of preventing infections that could lead to infertility, illness, even death. In menstruation, Ms. Profet suggests, the body takes a two-pronged attack against potential immigrants: it sloughs off the lining of the uterus, where the pathogens are likely to be malingering, and it bathes the area in blood, which carries immune cells to destroy the microbes. By these means, the pathogens and their home are destroyed simultaneously.

Ms. Profet's theory seeks to answer the simple question of why the bodies of premenopausal women go to the trouble of shedding considerable quantities of blood and tissue each month, losing iron and other valuable nutrients in the process. Why not keep the uterine lining around until an embryo needs it — why throw out the bath water if the baby hasn't been washed yet? And even if some of the lining must be turned over, why the copious bleeding? After all, the lining of the digestive tract is regenerated every two to four days, the skin sheds tens of thousands of cells every day, and other organs are freshened and patched up, all without the assistance of blood. In sum, menstruation is a costly event to the female, and Ms. Profet proposes that it wouldn't happen if it didn't serve an important purpose. She also believes that other types of uterine bleeding, like that which sometimes accompanies ovulation, the implantation of the embryo, and that which occurs after birth, may be the body's way of intermittently cleaning house and purging pathogenic intruders.

Going further, the theory holds that we and other higher primates are not the only mammals to menstruate, as is commonly supposed. Through an extensive review of scientific literature, dating back to the last century, Ms. Profet discovered that a number of mammals widely separated in evolutionary time have been observed to menstruate, including bats, marsupial cats, tree shrews, and primitive monkeys. If researchers only take the time to look, they may find that nearly all mammals menstruate, although many species may bleed trace amounts that escape easy detection.

This bold hypothesis has a number of medical implications. If bleeding helps to prevent infections, then women should avoid oral contraceptives that suppress menstruation entirely. In addition, inexplicable uterine bleeding should be viewed as a possible early sign of infection, a symptom that the body is struggling to thwart disease. Often doctors regard such bleeding as the result of abnormal hormonal flux, seeing it as a reaction that in turn increases a woman's risk of contracting a pelvic infection. But this attitude is completely backward, Ms. Profet insists, rather like saying a fireman causes a fire. If she is right, then the worst thing a doctor could do for an episode of unexplained uterine bleeding is to block the bleeding with hormones. A more appropriate response might be to test for an infectious organism like chlamydia and then prescribe an immediate course of antibiotics. There can be other reasons for unexplained bleeding — tumors, fibroid disease, or ectopic pregnancy — but infection should be considered as well, Ms. Profet argues.

The hypothesis also may explain the puzzle of why women who use intrauterine devices have extremely heavy periods. The IUD causes chronic inflammation of the uterus, and inflammation is generally a sign of infection. Reacting as if there are microbes about, the uterus increases the blood flow.

A number of gynecologists have attacked the theory, some because it's too novel for comfort, others because they say the women who come to them are more vulnerable to infection during their periods, not less. To which Ms. Profet responds (a) the moment of menstruation is not necessarily a time of stepped-up resistance but merely the rooting out of pathogens past, and (b) no immune mechanism in the body is perfect, and doctors usually see patients in whom the body's native defenses have failed.

The creator of the theory lives as well as thinks outside the mainstream. Ms. Profet received a strong imprimatur of her worthiness in 1993, winning, at the age of thirty-five, a MacArthur Award — the "genius" prize — but she has few other traditional credentials. She never bothered to get a doctorate, viewing it as a waste of time and a potential damper on creativity. Instead, she has published unorthodox theories on the evolution of commonplace phenomena that scientists and physicians have generally ignored. For example, she has proposed that morning sickness, long thought to be an incidental aspect of pregnancy, in fact evolved to prevent women from eating vegetables and other foods that are rich in natural toxins at a time when the developing fetus is most vulnerable to ingested poisons. She has also suggested that some people suffer from allergies as a way of protecting themselves against plant-borne compounds that would damage their cells if not expelled from the body by a sneeze or cough.

Ms. Profet first considered the problem of menstruation when she learned of it, at the age of seven, from an older sister. "I was disgusted because it made no sense and seemed so inefficient," she said. "Why go to all that trouble to make that elaborate lining just to get rid of it? I thought, God must really hate us to come up with something so ridiculous." As she grew older, the clinical explanations of menstruation dissatisfied her, and she was irritated by medical descriptions of a woman's period as the unfortunate and possibly unnecessary byproduct of hormonal cycling, theories for which, as far as she could tell, there was no proof.

Just as the insight into the structure of benzene came to its originator in a dream of a circle of snakes biting one another's tails, so Ms. Profet did her most imaginative thinking in her sleep. One night she dreamed of black triangles stuck in deep red tissue, and, on waking, realized that the triangles represented pathogens, the scarlet backdrop a bleeding womb.

Through considerable research, Profet came up with converging lines of evidence to support the theory. She first sought proof that menstruation is an adaptation, something that has evolved to meet a defined goal, rather than being a meaningless side effect of oscillating hormones. By studying physiology, she learned that there are distinctive blood vessels, the spiral arteries, that open to the uterus and orchestrate menstruation by first sharply closing and then rapidly dilating. The closing of the arteries kills the tissue by depriving it of blood, and the reopening lets in a rush of blood that forces off the newly necrotic tissue. In addition, menstrual blood is notably lacking in clotting factors, which cause blood elsewhere in the body to coagulate on exposure to air.

Satisfied with the evidence showing an adaptive design to menstruation, Ms. Profet pursued her theory of what bleeding is designed for. She found abundant clinical confirmation that sperm is a potent vector of disease; electron micrographs of sperm invariably show the familiar teardrop cells girdled with tag-along bacteria. And though mucus around the cervix generally inhibits the passage of any organism into the upper reproductive tract, the mucus becomes permeable during ovulation, when sperm must be permitted access to a waiting egg. By piggybacking on sperm, microbes originating with the male or picked up from the vaginal canal during intercourse can glide through the cervix and invade a woman's organs, putting her or any embryo at risk of disease.

Other evidence points to the period's protective role. Menstrual blood is rich in macrophages, immune cells that engulf infiltrators, and is able to sequester iron and keep it from bacteria, which require iron to survive.

Knowing that other female mammals would be exposed to sperm-borne microbes as readily as women are, Ms. Profet sought proof that menstruation and other types of uterine bleeding are common in the mammalian kingdom. She came up with a list of many species that bleed either visibly or covertly, and could find no persuasive evidence that any female does *not* menstruate. That human periods are the most noticeable of all is hardly surprising; women are sexually receptive more often than any other female mammal and thus are at greatest risk of uterine infections from sex. Women are freed from the need for bleeding during pregnancy, when the cervix is fairly well sealed off from sperm by a thick and chemically hostile coat of mucus. (During the last two months of pregnancy, however, the mucus becomes more permeable, and some doctors advise women to have their partners use condoms to protect against sperm-borne infections.) Postmenopausal women also have thicker cervical mucus than do fertile women; it presents a barrier that at least partly offsets the loss of the body's monthly housecleaning. When sperm cells need no longer be permitted entry to find a receptive egg, they and their microbes may as well be blocked at the cervical gate.

29

WHY VEGETABLES ARE
GOOD FOR YOU

CACHED AWAY in the soul of every red-blooded American who fondly recalls when carnivory was a virtue and supper wasn't supper without a centerpiece of pork chops or prime ribs lies the frail hope that all the recent emphasis on fruits, grains, and vegetables, vegetables, vegetables will somehow turn out to be a terrible mistake.

Abandon that hope, ye who succor it. The truth is that the more we learn about the ingredients found in fruits, vegetables, beans, and herbs, the more impressive appears the power of those compounds to retard the bodily breakdown that results in cancer and other chronic diseases. Nutritionists and epidemiologists have long observed that people who eat a plant-rich diet suffer lower rates of cancer than do meat loyalists, and now scientists are on their way to understanding why.

Beyond bearing the benefits of vitamins and fiber, plant foods are rich with chemicals that have no nutritional value and are not necessary for immediate survival yet may impede cancer at several stages in its slow, savage evolution. Most of the experiments performed so far have been done on animals or isolated cells, and no specific ingredient from fruits or vegetables has been proved in long-term human trials to prevent or arrest malignant growth. Nevertheless, we can find encouragement (or disgruntlement) in the harmony of laboratory results with the empirical studies of long-lived populations.

Just when researchers thought they had a reasonable grasp of the basic anticancer compounds found in a healthy diet, they discovered a novel pathway through which ingredients in plants may help foil disease. Scientists at Children's University Hospital in Heidelberg, Germany, recently isolated a compound called genistein from the urine of people who eat a traditional Japanese diet, heavy on soybeans and vegetables. Through petri dish experiments with a synthetic version of the chemical, they demonstrated that genistein blocks the event called angiogenesis, the growth of new blood vessels.

This talent may have significant implications for both the prevention and treatment of many types of solid tumors, including malignancies of the breast, prostate, and brain. When a tumor seeks to expand in size beyond a millimeter or two — about the dimensions of a hyphen on this page — it first must foster the growth of new capillaries around it. Once it is fully vascularized, the malignancy receives the oxygen and nourishment it needs to keep swelling, and it eventually invades the blood and lymph system and seeds fatal metastatic colonies elsewhere in the body. By inhibiting capillary growth, genistein just may keep nascent tumors from growing beyond harmless dimensions.

Genistein is found in high concentrations in soybeans and to a somewhat lesser degree in other leguminous plants. In those on a traditional Japanese diet, the urine level of genistein is at least thirty times that of Westerners. Such a diet could explain why, when Japanese men leave their country for several years to work in the United States or Europe, their rate of invasive prostate cancer rises sharply. Any tiny prostate tumors that had been kept in check by daily intake of, say, miso soup finally are free to grow once the men assume a more Western, genistein-poor culinary style. If genistein proves its mettle through testing in animals and controlled clinical trials, the compound may be useful both as a dietary measure to prevent cancer and, in concentrated form, to treat tumors already in progress.

Blocking angiogenesis is considered an ideal sort of therapy, one that can attack the malignancy while leaving normal tissue intact. Other than meeting the sinister demands of tumors, new blood vessels grow in the adult body only after fairly rare events like severe injury, heart attack, or the implantation of an embryo in the uterus; thus, any compound that impedes angiogenesis would have few side effects.

Encouraging as the findings of anticarcinogens in foods may be, researchers admit that the field of food analysis is in its infancy. Food is chemically daunting, with every stalk of broccoli or slice of melon composed of hundreds or thousands of individual yet interacting chemicals. Some plant products contain natural toxins that promote cancer along with compounds that inhibit the disease, and it can be difficult to sort out which class of chemicals predominates in a given food. Beyond its inherent difficulties, the nutrition business is prone to faddishness, charlatanism, and hype, peopled as it is with fanatical adherents of vitamin supplements, the anti-aging crowd, strict frugivores, immoderate legumivores, and the like. Mainstream researchers have tended to shy away from it.

Nor has there been much encouragement for studies focused on the prevention of cancer rather than its treatment. On average, only about 5 percent of the approximately $2 billion annual budget of the National Cancer Institute is earmarked for disease prevention; far more goes toward expensive and high-profile studies like those on gene therapy, which, if it works, will take years before it is of use to many cancer patients. The cost of bringing the widely touted drug taxol to market is estimated at $1 billion, yet taxol adds only about five months to the life of an ovarian cancer patient. Would that a similar sum were spent on preventing the cancer in the first place, a task that demands greater knowledge of the things people put in their mouths.

To be sure, some of the benefits of a plant-based diet are what it helps one avoid: a person taking in lots of fruits and vegetables is less likely to fill up on fatty foods. Moreover, vegetables have fewer calories than do meats and cheeses, and restricting calorie intake has been shown in animal studies to sharply reduce the incidence of cancer. Yet apart from the virtues of omission, the positive benefits of vegetables are many. In the course of metabolizing energy and using oxygen, the body's cells constantly generate hazardous molecules called free radicals, which can mutate genes and set the foundation for cancer. Most of the radicals are sopped up by the body's native antioxidant enzymes, but yellow and green vegetables, as well as melons and citrus fruits, also offer a wealth of antioxidant compounds, including vitamins C, E, and beta carotene, the precursor to vitamin A. Animal experiments have shown that rosemary, green tea, and curcumin — the chemical responsible for curry's yellow pigment — all suppress cancer growth, very likely by acting as antioxidants and neutralizing free radicals before they reach the cell's kingpin DNA.

Scientists have explored the influence of plant chemicals on estrogen metabolism and the ways that diet may inhibit breast cancer. It's known that estradiol, the precursor to estrogen, takes one of two metabolic pathways, turning into either a 16-hydroxylated or a 2-hydroxylated form of estrogen. The 16 form is stimulatory and has all the earmarks of being a comparatively dangerous version: women with a high risk of breast cancer show elevated levels of the 16 type in their blood, and tissue from breast tumors contains more of the 16-hydroxylated form of estrogen than does surrounding, noncancerous breast tissue.

By contrast, the 2-hydroxylated form is relatively inert and has been found to be elevated in women who are vigorous athletes and thus have a lower than average risk of breast cancer. Of relevance to our story, the inactive type of estrogen also predominates in women who eat many cruciferous vegetables like broccoli, Brussels sprouts, and cabbage.

By isolating ingredients from these leafy green vegetables, researchers have shown that one chemical in particular, indole-3-carbinol, can induce estradiol to follow the harmless metabolic route toward 2-hydroxylation. To see whether that inducement makes any difference to women's cancer risk, in mid-1993 the researchers began giving a group of women daily capsules of 400 milligrams of indole carbinol, equivalent to the amount in half a head of cabbage. Within the first few weeks of the study, the participants' blood levels of the harmless 2-hydroxylated estrogen had risen to concentrations seen in marathon runners and remained elevated for months; but whether the change in estrogen metabolism affects breast cancer rates will take years to sort out.

The cruciferous vegetables are a motherlode of anticancer compounds, each chemical displaying its own method against cell madness. Another protective ingredient is sulforaphane, the most robust member of a chemical class known as isocyothionates, which lend broccoli, cauliflower, kale, mustard, horseradish, and many other vegetables and spices their tangy snap. The isocyothionates seem to guard against cancer indirectly, by stimulating the body's production of naturally occurring protective enzymes — phase 2 enzymes — which latch on to carcinogens, detoxify them, and swiftly flush them from the body.

Scientists have been able to identify sulforaphane and other enzyme inducers in foods through a simple system that relies on cultured mouse liver cells and scans for a spike in phase 2 enzyme activity. Applying the screening method not only to different vegetables, but to diverse varieties of the same vegetable, they have found that the amount of inducer ingredients varies tremendously from one sample to the next, either because of natural genetic variations among strains, or because of the different methods in vegetable cultivation.

Nothing in food analysis is proving to be simple, and everything appears to do many things. Vitamin C, beyond its antioxidant muscle, also inhibits in the stomach the creation of nitrosamine, a potentially dangerous carcinogen. In addition to suppressing blood vessel growth, genistein acts directly on can-

cer cells and deters their proliferation. Fiber, which almost sin-
glehandedly resuscitated the breakfast cereal industry (but also
is at the heart of fruits and vegetables), exerts an array of positive
effects on the body. It dilutes the concentration of noxious compounds in the colon so that the toxins have less chance of
harming the delicate mucosal tissue there, and it moves everything through the system faster. Fiber also alters the environment
of the gut and colon. Through a poorly understood mechanism,
it discourages the growth of harmful bacteria that release enzymes believed to promote cancer by transforming precarcino-
genic chemicals in food into active agents of malignancy. At the
same time that it hinders undesirable microbial expansion, fiber
bolsters the growth of benign bacteria, which further crowd out
the unsavory strains. As though that weren't reason enough to
trade chewing the fat for chewing the cud, fiber also encourages
the creation of the healthier form of estrogen, and thus may
impede breast cancer.

Considered together, the intricacy and synchronicity of the
chemicals in plants argue firmly against an undue reliance on
vitamin supplements to compensate for a rotten diet of snacks
and french fries. If scientists have yet to understand all the
subtleties of a Brussels sprout, how can anybody hope to reca-
pitulate it in a pill?

30

UNSIGHTLY FAT:

THE MAMMALS' FATE

FEW THINGS IN LIFE are as vexing as the talent that fat has for gathering in unsightly pockets and paunches at certain spots around the body: jiggling on the thighs, bulging out from the belly, flapping down from the triceps like bedsheets drying in the wind. But distressing though such fat deposits are, they may simply be the price we pay for being mammals.

Researchers once thought that the body's fat was distributed under the skin and around internal organs in a fairly uniform fashion, and that people who became pudgy at one or another location did so as a result of their genetic peculiarities or exercise habits. However, as a rash of studies comparing human fat tissue with that of other mammals makes distressingly plain, even the leanest wild mammals cannot help putting on fat in a few discrete hot spots around the body — just those places we bemoan for their seemingly infinite storage capacity. Whether you're talking about a squirrel, a badger, a deer, a wolverine, a camel, or a person, the fat seats itself in the breast area, around the upper part of the front legs (our upper arms), on the tailbone and around the thighs, in three to eight regions of the abdomen, and at the back of the neck. In many mammals, a surprisingly sizable glob of fat also surrounds the heart, a discovery that contradicts conventional notions that fat near the heart is a pathological condition largely confined to humans. The total amount of body fat varies from species to species and individual to individual, but when fat congregates, it is always at these choice addresses.

Location, it seems, is everything. The new work on the biology of fat — called adipose tissue by those who study it in the laboratory, lard by those who study it in the mirror — reveals that fat cells display notably different biochemical properties depending on where they are in the body. Some are efficient at absorbing lipids, or fat molecules, from the bloodstream, while the cells in other deposits are primed to release lipids easily as fuel for neighboring tissue. Indeed, the various fat deposits on the thighs, belly, and around the viscera may all be thought of as substantially different organs.

Behind the accumulation and deposition of adipose pucker are enzymes that synthesize, process, and store fat molecules. Of interest is an enzyme called lipoprotein lipase, which plays a leading role in extracting fatty acids after a meal and storing them in fat cells. Lipoprotein lipase has been detected in nearly every species examined, and it is more abundant in females than in males, probably to allow females to store fat easily for pregnancy.

The exquisite regulation of that enzyme and others may explain why bears and woodchucks and such animals that become stout each year before hibernating or fasting do not suffer the ill effects of obesity often seen in humans, like high blood pressure, clogged arteries, and diabetes. A polar bear, for example, can eat so much seal blubber that the fat in its blood would instantly kill a dog and possibly even you; yet a polar bear's liver and arteries are comparatively fat-free, and it doesn't get heart attacks. The reason appears to lie in the performance of its lipoprotein lipase. In the dreamiest of all possibilities, the new understanding of fat metabolism will suggest better treatments for obesity. At the very least, scientists hope to convince people that all fat is not created equal, nor is all fat equally bad.

Whatever its biochemical peculiarities, the bottom line on fat is that it serves as a handy store of energy in hard times. Fat in the diet is almost effortlessly converted into fat on the body. In translating consumed fat into storable fat, the digestive system spends only 2 percent of the fat molecules' inherent energy, letting the rest get dropped off in fat cells. By contrast, the body burns half the calories of consumed carbohydrates in the metabolic operation required to transform starch into storable energy. In other words, it's much easier to make fat from fat than it is to make fat from any other food item. Moreover, the adipose tissue that stores fat molecules for future use can expand almost indefinitely, an unusual property shared by no other organ save the skin. Part of that flexibility results from the nature of the fat cells that make up adipose tissue.

Individual fat cells can balloon up to ten times or more their original size. A fat cell is just a membrane encasing a big round fat droplet, and the membrane can be stuffed with more fat rather as a sausage casing can be stuffed with pork. After a while, if the fat in your diet proves too great for existing fat cells to sop up, your body will propagate new ones, and those fat cells never die. When you lose weight, the fat cells shrink, but they lie in wait for the next fatty feasting.

It is because fat at first seemd so simple and amorphous that researchers long neglected to notice the significance of its accretion in certain regions of the body. But then epidemiologists began to see that people who tended to put on weight around their bellies were at higher risk for heart disease than were those who spread out in the thighs and buttocks. That observation spurred scientists to wonder about the biochemical profile of various fat deposits and to look at the portliness of our fellow mammals for clues. Some of the more intriguing findings have come from studies of the slab of fat that surrounds the heart. At first glance, the notion of fat deposits around the heart seems like a lousy bit of engineering, because the heavy adipose pad has to be moved every time the heart takes a beat. Yet the fat tissue here proves to be most accommodating. It has an exceptional ability both to take up fatty acids in the blood and to generate the lipids the heart muscle needs as fuel. It also acts as an absorbent buffer, shielding the delicate heart muscle from too much fat after an especially greasy meal.

Each fat deposit has its own biochemical charm. The bulges around the thighs are efficient at absorbing lipids from the bloodstream and reluctant to give up the fat treasure once it is stowed. At the same time, thigh fat is relatively poor at taking up glucose, simple sugar that provides immediate sources of energy. By comparison, the small amount of fat found between muscles in the body prefers to soak up glucose from the blood, which it turns into lipids to stoke the hungry muscle tissue beside it. The difference in biochemistry serves a purpose: the intramuscular fat is designed for fast response and quick storage; the thigh fat is fashioned to take care of long-term needs. These days, of course, most of us don't have long-term needs, but our thighs somehow haven't caught up with the era of twenty-four-hour supermarkets. There's some comfort to be had in saddlebags, though: after a fatty meal, they rapidly remove the fat from the bloodstream before it has a chance to cling to the arteries. From a health perspective, then, thigh fat is not a bad fat to bear.

The thighs are not the only ones that have it. As many women are all too aware, they are overall a bit plumper than men on average. Here, too, clues to the differences can be found by studying other creatures. By examining the enzymes involved in lipid metabolism throughout the mammalian kingdom, scientists found that lipoprotein lipase, the main enzyme responsible for storing fat, is partly controlled by reproductive hormones. Both males and females use the enzyme to help lay down fat deposits, but in the females of many species, female sex hormones somehow stimulate the rapid production of the lipase enzyme, allowing the females to fatten up either before or during pregnancy. In humans, the sex differences between lipoprotein lipase activity are more extensive still, and help explain discrepancies in the distribution of fat in men and women. In women, the fat cells of the hip, thigh, and breast region tend to produce and secrete the enzyme; in men, the fat cells of the stomach area are likelier to generate the lipase.

The pattern of lipase activity in men provides the mechanism for midriff weight gain, but why men's belly fat cells produce abundant amounts of the enzyme in the first place remains unclear. The trait may have arisen early in human evolution, perhaps so that men could assume the role of hunters, warriors, and, on occasion, fleers. The fat around the belly and abdominal organs is highly responsive to the famed fight-or-flight stress hormones; when tweaked by the hormones, the abdominal fat cells readily release fatty acid fuel for quick use by the muscles and heart. But whatever its initial virile purpose, belly fat in modern man mostly presents a health hazard, especially when it is ignited by the chronic stress that animates our post-advanced civilization. With the constant stimulation of adrenaline, the fat cells continuously release fatty acids into the bloodstream. Because of the design of the circulatory system, the freed fatty acids go directly to the liver before being distributed throughout the body for use by the muscles. When too many fatty acids drain into the liver, the organ becomes resistant to insulin. As a result, the blood glucose levels soar and the pancreas produces more insulin, which can lead to high blood pressure, diabetes, and eventual heart disease.

The peril is not confined to men. They may be more prone than women to developing belly fat, but the women who tend toward apple-shaped bodies run risks of cardiovascular problems comparable to those of paunchy men. By the same token, men who have the pear-shaped figures often seen on women, with their fleshly emphasis on thighs and buttocks, have a comparatively low risk of heart disease.

The most optimistic researchers believe that new treatments for battling pathological human obesity may come from understanding other mammals that manage their fat with aplomb. As a rule, we are more prone than wild animals to getting fat, simply because those of us in developed nations have both constant access to food and a metabolic system that anticipates occasional famines. But other creatures do get fat in the forest, and when they do, they do it in high style. Among a free-foraging population of eleven hundred macaque monkeys living on the island of Cayo Santiago, off the southeast coast of Puerto Rico, about 6 percent end up obese, bulking to almost twice the standard macaque weight of twenty pounds. Yet not one of the mammoth monkeys has ever been found to develop obesity-related diseases like high blood pressure or diabetes, and they remain active and fully fertile throughout their lives. There's a more spectacular example of fatness in the wild: the woodchuck. Every year the woodchuck trebles or quadruples its weight to make it through winter, yet the animal remains spritely, with not an atherosclerotic plaque to be found. Seeking to learn why the animal becomes so big so beautifully, researchers found that just as the creature is preparing to hibernate, its levels of lipoprotein lipase and other enzymes related to lipid metabolism soar, the better to lay on the fat as rapidly as possible. By spring, however, the enzyme levels have dropped to normal. That is a stark contrast to the biochemical dynamics in humans. People who have been obese for a long time retain elevated levels of LPL and related enzymes even after they lose weight. Why the enzymes stay high in humans remains unclear.

Other mammals may also have something to teach us about the chemistry of will power. At some point in the fall, for example, bears getting ready for the big sleep decide to gain enormous quantities of weight by eating enormous quantities of fish. In the spring and summer, the bears decide to stay slim, eating fewer fish despite any ambient riches of lake and stream. They just don't want the extra pounds to weigh them down and heat them up when the weather is warm. Something in the brain, some chemical cue, lends them that easy springtime self-control, and if that something could be isolated and synthesized, it would be one product we all might greet with the happiest of bear hugs.

31

THE ANATOMY

OF JOY

WE ARE A NATION born of puritanism and bottle-fed on priggishness, so perhaps it is no surprise that science has applied far stronger zeal to the anatomy of melancholy than to the understanding of unfettered joy. Researchers have analyzed the stress response in exquisite detail. They know that perpetual surges of adrenaline, noradrenaline, and cortisol sear the body like a drizzle of acid, and that chronic stress, along with its loyal comrades, anger and depression, can sicken and even kill us. They suspect, however, that sensations like optimism, curiosity, and rapture — the giddy, goofy desire to throw the arms wide and serenade the sweetness of spring — not only make life worth living but also make life last longer. They think that euphoria, unrelated to any ingested substance, is good for the body, that laughter is protective against the corrosive impact of stress, and that joyful people tend to outlive their bilious, whining counterparts. Researchers who followed the fortunes of a group of medical students over twenty-five years found that, among those who rated on the easygoing end of a test that measured temperament, only 2 percent died by the age of fifty. By contrast 14 percent of those who scored as churlish and hostile died by that age.

Why happiness is healthy, though, and what the body is doing when it exults in itself, science has only the most suggestive of clues. Endorphins, the brain's built-in version of opium and the chemicals that reputedly account for the runner's high, apparently have less to do with joy than with easing the perception of pain. Another compound that gets attention these days is oxytocin, the small hormone secreted by the pituitary gland and the possible mediator of feelings of satisfaction and harmony (see Chapter 2). But most experiments thus far have been limited to rodents, and scholars of behavioral science seem reluctant to hunt down the human version of a cuddle drug. Doctors prefer to focus on serious things — that is, the things that make people sick.

Sadly, too many scientists describe happiness from a negative perspective. By this logic, happiness is regarded as healthy only because it spares us the enfeebling impact of anxiety or inspires us to cultivate such worthy habits as eating vegetables, avoiding liquor and cigarettes, and sleeping eight hours a night. Yet even the most grim-faced researcher knows that real joy, far from being merely a lack of stress, has its own decidedly active state of possession, the ripe and gorgeous feeling that we are among the blessed celebrants of life. It is a delicious, as opposed to a vicious, spiral of emotions.

Researchers complain that one reason they have trouble understanding happiness is the difficulty of recapitulating this emotion in the laboratory. You can make people angry by having them sit in your waiting room for a couple of hours, you can make them anxious by telling them you've found a disturbing lump, but it's almost impossible to make them happy short of giving them, say, a line of cocaine, in which case you've defeated the purpose of seeing what happens in a natural state of cheer.

One version of joy that does yield to laboratory analysis is laughter. Sustained hilarity, it turns out, is among the more agreeable forms of aerobics. While you're laughing, the muscles of your abdomen, neck, and shoulders rapidly tighten and relax; heart rate and blood pressure both increase; inhalation and expiration become spasmodic and deeper. When the laughter subsides, your blood pressure and pulse are likely to fall to lower, more salubrious levels than before the merriment began. A hundred laughs are the equivalent of ten minutes of rowing, the difference being that you're actually able to smile while you're laughing.

Laughter also helps gird against discomfort. In one experiment, students were shown a videotape of Bill Cosby (performing as a standup comic, not as a saintly primetime father) while a control group viewed an instructional videotape on the fine art of hanging plants in baskets. Both groups received mild but increasingly insistent electrical shocks and were told to signal when the pain grew too great to bear. Perhaps to nobody's astonishment, the Cosby watchers were able to endure a significantly higher jolt than the plant viewers could — though how much of the latter group's agony arose from the tedium of their enforced entertainment cannot be vouched for.

Even when studying jocularity, researchers cannot prevent themselves from considering its downside. Among the relatively scant listings on *Laughter* in the scientific literature we find an inordinate number on "pathological laughter," cases of psychiatric patients who laugh to mask miseries ("Treating Those Who Fail to Take Themselves Seriously," as a New York psychiatrist put it) or of brain-damaged patients who laugh for no reason. One paper describes the case of a man who arrived at an emergency room in Ohio after having accidentally inhaled a mild insecticide. He had no symptoms beyond slight numbness, tremors, and uncontrollable laughter. Doctors could find no physical or neurological damage, but the man continued to laugh for fifty-five minutes, so long that he complained his abdominal muscles were killing him. He was given a tranquilizer intravenously, his laughter ceased, and the doctors sent him home — no doubt with frowns of triumph plastered firmly on their faces.

VI

CREATING

32

THE ARTFUL

DOCTOR

WHETHER IN LIFE or after death, Vincent van Gogh has never been a painter of moderation. He painted ferociously, drank great quantities of the potent liqueur absinthe, went for days without eating, slashed off his left ear lobe, and committed suicide at the age of thirty-seven. He sold only one painting in his lifetime, but today his works command tens of millions of dollars. And since his suicide, in 1890, the great postimpressionist has been the subject of no fewer than 152 posthumous medical diagnoses. Doctors poring over van Gogh's paintings and his extensive correspondence have variously claimed that he suffered temporal lobe epilepsy, a brain tumor, glaucoma, cataracts, manic depression, schizophrenia, magnesium deficiency, and poisoning by digitalis, which once was given as a treatment for epilepsy and can cause yellow vision — thus explaining, the story goes, van Gogh's penchant for brilliant yellows.

Most recently, *The Journal of the American Medical Association* offered two original if suspiciously neat diagnoses of van Gogh's epic afflictions. One posited that he suffered from Ménière's disease, an inner-ear disorder that causes vertigo and may have prompted the artist to slice off part of his ear; the other argued that the Dutch master had acute intermittent porphyria, a hereditary metabolic disorder capable of causing hallucinations, derangement, depression, seizures, abdominal cramps, and other symptoms van Gogh often complained of, and that just so happens to be exacerbated by fasting, alcohol binges, and exposure to paint fumes.

The latest entries in the van Gogh malaise-of-the-month club are part of a continuing exercise that certain aesthetically minded doctors engage in, either for cerebral sport or for a better understanding of the natural history of diseases. The game is called "Diagnosing the Canvas." In one approach, physicians attempt to identify an artist's illness or to chart its progression by considering suggestive details in the artist's work, like color choice, perspective, and subject matter. That sort of analysis has yielded the proposal that Claude Monet's near-blinding cataracts and eventual eye surgery deeply influenced the evolution of his water lily series, and that Goya's depiction of paranoid and emotionally volatile figures reflected his feelings about his worsening deafness.

In the second version of the pastime, doctors study abnormal or deformed subjects portrayed in works of art and attempt to explain a figure's anomalous appearance by making a medical diagnosis. Noting the distinctively gnarled hand of the woman shown in Corot's painting *Girl with Mandolin,* two physicians and an art student have suggested that the musician had rheumatoid arthritis, a crippling autoimmune condition relatively common among young women. They further argue that Corot painted the hand in such a twisted manner because he himself suffered from gout, another arthritic condition, and perhaps was obsessed with the symptoms of the disorder.

On occasion, the artistic depictions of disease enable medical historians to estimate when an illness first penetrated a particular population, and how widespread it may have been. Rheumatoid arthritis, for example, is a genetically linked disease. Observing that the deformity is not depicted in any European work of art before 1800, rheumatologists theorize that the genetic mutation may have been absent or rare among Europeans before that point. The anonymous sculptors and painters of ancient Egypt and Central America represented a number of hereditary diseases with such anatomical accuracy that they are used today to teach medical students. Among the outstanding examples is the first known depiction of retinoblastoma, an inherited eye cancer that, as a Mayan sculptor showed all too graphically, results at its late stages in a tumorous mass bulging forth from the eye socket.

Most art-loving doctors say they engage in diagnosing canvases less for scientific reasons that because it is an irresistible diversion. They feel a kinship with artists, since a good diagnostician, like a good painter, observes the tiny, revealing details that ordinary eyes usually miss. In diagnosing an illness, a doctor looks for a small discoloration on the face, the pit in the fingernail, the superficially dilated capillaries on the skin. So when doctors see artists paying attention to the same clues in their paintings, they can't help regarding the canvas as a patient, silently awaiting their professional opinion.

For all the pleasure it affords, however, the temptation to pin a syndrome to an artist the doctor has never met, or to diagnose a painted figure unable to so much as say where it hurts, has led to outlandish and even harmful notions about art and artists. As modern art became less representational, for example, people tried to explain it away as a product of disturbed minds at work, and they solicited the opinions of physicians and psychiatrists to confirm that, yes, this or that artist had a real mental disability. Art and medical historians are still trying to debunk once and for all an especially famous effort to define an artist's distinctive style through a pat diagnosis. In 1913, Parisian doctors suggested that El Greco painted his elongated figures because he may have had astigmatism, a vision problem in which the eyeball is shaped more like a football than a sphere. In some types of astigmatism that have been corrected with glasses, objects may appear slightly elongated in one direction and squashed in the other.

But as ophthalmologists and others have repeatedly argued in the intervening years, the theory about El Greco is nonsense. To begin with, an astigmatic whose vision is not corrected with glasses doesn't see objects as elongated, but merely as blurs, and there were no corrective lenses for astigmatism in El Greco's day. In any case, X-ray images taken of El Greco's paintings show that beneath the painted figures are drawings of a more naturalistic composition, indicating that the artist consciously chose to stretch out his images when he applied paint, very likely to lend them an ethereal quality.

Medical canvassers are on safer ground when they consider an artist's work in the context of contemporaneous medical documentation. In one such study, doctors considered how the progress of Monet's cataracts affected the evolution of the great impressionist's art. Monet's condition was diagnosed in 1912, when he was seventy-two; but considering its gradual and insidious nature, the disorder almost surely began many years earlier. Toward the end of the nineteenth century, his paintings grew fuzzier and muddier, with ever fewer details. The colors became yellowish-brown in cast, the color range most visible to those with cataracts; least visible are the blues and violets. Monet himself described the visual changes to a reporter in 1918: "I no longer painted light with the same accuracy. Reds appeared muddy to me, pinks insipid, and the intermediate or lower tones escaped me."

By 1922, Monet was legally blind, able to see light but almost no form or color. After he had cataract surgery in his right eye, his ability to see blue colors returned so sharply that he couldn't stand the brightness and had to wear glasses with a yellow tint. In the last four years of his life, Monet completed his water lily cycle, and some of those paintings glow with soft, lush blues and lavenders.

Edgar Degas may have suffered late in life from another eye disease, macular degeneration, in which the center of the retina gradually degrades. A person with the condition retains peripheral vision but loses central vision, and Degas's later paintings emphasize life, form, and motion at the edges, while the middle of the canvas appears unfocused, almost irrelevant.

.   .   .

Doctors also delight in analyzing paintings that portray their own profession. The American realist Thomas Eakins, who studied medicine before turning to art, painted a famous scene of an operating clinic where surgeons are removing a piece of bone from a boy's thigh while the boy's mother sits nearby. Examining the nature of the problem, doctors have proposed that the boy suffered from myelitis, an inflammation of the bone marrow that causes pieces of bone to separate and become sequestered in pockets within the leg. Since the painting was done early in the century, the underlying cause of the myelitis was likely to have been tuberculosis.

Perhaps the most famous doctor painting is by the English artist Sir Luke Fildes, called, simply, *The Doctor.* A physician sitting on a chair watches a child who is sleeping across two chairs while the parents, in the background, look on. Although he clearly can't do much for his young patient, the doctor looks vaguely hopeful. Moreover, light is just beginning to filter through the window, a possible visual metaphor to show that the worst of the child's illness has passed. What the disease may have been is unknown, but some speculate that it was a sudden childhood infection like scarlet fever or lobar pneumonia.

Many artists without medical training were fascinated by disease and deformity. Velázquez was renowned for painting dwarfs, cripples, and children with obvious birth defects. Edvard Munch, the Norwegian expressionist whose sister died of tuberculosis, painted a number of portraits of the ailing, among them a picture of a good friend's sister. The girl — wrapped in a blanket, her face flushed, her hand clutching a single flower — appears to be suffering from an illness, perhaps tuberculosis, which was then prevalent throughout Scandinavia. Art historians have long debated the significance of the flower; some see it as a symbol of her frailty and an indication that she will soon die, while others consider it a sign of hope.

Yet for every instance of a clever diagnosis, a physician is as likely to neglect the larger context in which a masterpiece reigns sublime. Doctors have been known to complain about Michelan-gelo's *Pietà* in Rome for its supposed anatomical inaccuracies. Look at Christ's arms, they say. He's supposed to be dead, but the veins of his arms and hands are engorged as though with blood. Everybody should know, they continue, that when the heart stops, the surface veins almost immediately flatten out. The criticism enrages art historians, who are aware that Michelan-gelo painstakingly studied and dissected cadavers to master the subtleties of the human body. And they are equally aware that Michelangelo had no interest, as he carved the *Pietà,* in depicting Jesus as an ordinary corpse. Instead, one must observe the telling details, the story of the Resurrection nested within the story of the Crucifixion. Christ's dropped right arm fingers the hem of the Virgin's garment, a gesture of love like that of a child. His other hand is clearly gesturing, and, yes, the veins of his arms are engorged with blood. The diagnosis: from death will spring life, and humanity's deepest sorrow holds the seeds of its ulti-mate joy.

33

FROM MADNESS
TO MASTERPIECE

As LONG AS there have been poets to pierce the darkness with their diamond songs, and painters to capture blades of sun shattering on cool cathedral stone, and artists of all persuasions to consort with the gods and articulate the union, there have been social critics to notice that an awful lot of these creative types are mentally unsound.

"Why is it," Aristotle asked in the fourth century B.C., "that all men who are outstanding in philosophy, poetry, or the arts are melancholic?" Two thousand years later, the English poet John Dryden wrote, "Great wits are sure to madness near allied; And thin partitions do their bounds divide," a sweet couplet that has since degenerated into the sorry cliché "There is a thin line between genius and madness." Yet, as with any cliché worth the iteration, this one holds a sizable grain of truth. After many decades of quarreling over the definitions of slippery and subjective terms like *madness* and *creativity,* along with a general resistance among scientists to any idea that has gripped the popular imagination for so long, psychiatrists, neurologists, and evolutionary geneticists have accrued powerful evidence that the link between mental disturbance and artistic achievement is real. Study after study has shown that people in the arts suffer disproportionately high rates of mood disorders, particularly manic-depression and major depression.

Those with manic-depression, or bipolar disorder, oscillate between summit and abyss — between a sense of grandeur and recklessness, an unbounded knockabout energy that feasts on itself and disdains the need for sleep; and a profound depression in which anguish, lethargy, and self-hatred dominate. Many of the most eminent creators seem to have had full-blown manic-depression, others to have had milder forms of the disorder, while still others have suffered repeated episodes of major depression, the same bleakness seen in the downswing of manic-depression but without its euphoric counterpart.

The list of artists in whom manic-depression or severe depression has been diagnosed with confidence is a pantheon: Lord Byron, Percy Bysshe Shelley, Herman Melville, Robert Schumann, Virginia Woolf, Samuel Taylor Coleridge, Ernest Hemingway, Robert Lowell, Theodore Roethke, to name but a very few. Depending on which study you look at, the most outstandingly creative individuals suffer from bipolar disorder and depression at rates ten to thirty times that found in the general population. And though creativity is an essential element in many professions, the link between creativity and mental instability is more pronounced in the arts than in other fields. In a study of 1004 prominent individuals of the stature of Aldous Huxley, Alexander Graham Bell, Albert Einstein, and Henri Matisse, psychiatrists found that psychiatric disturbances were comparatively more common among artists. For example, the rate of alcoholism was 60 percent among actors and 41 percent among novelists, but only 3 percent among people in the physical sciences and 10 percent among military officers. In the case of manic-depression, 17 percent of the actors and 13 percent of the poets suffered from the disorder, while the incidence was less than 1 percent among scientists, about the same as the rate in the general population.

None of this is an attempt to romanticize mental illness or to argue that you've got to be crazy to be great at your art, nor, conversely, that the poems of your average madwoman are any likelier to be immortal than those of your average insurance salesman. The anguish and dangers of mood disorders remain extreme: if not treated with lithium or other medications, 20 percent of patients with manic-depression will commit suicide. And of course artistic achievement requires sustained effort and personal sacrifice, a dedication beyond what most mortals are capable of no matter how many neuroses they can claim.

Indeed, a hallmark of all accomplished artists who happen to be mad is that their bouts of mania or depression are interrupted by long stretches of normality, in which the creators are in utter command of their work. During these healthy periods, they are better than well, stronger than most, more confident and focused, instruments tuned to celestial scales. They work brilliantly and productively while they are sane, but when the blackness returns, they plunge again into paralysis. They are both scarred and whole, diseased and robust.

Yet while the interludes of health are essential to creative production, the moments of madness may add their own notes of brilliance. The neurobiology of a mood disorder may work in profound ways to nourish or whet creative thinking, and perhaps there is a link between mental instability and inspiration. Because manic-depressives are ever riding the biochemical express between emotional extremes, their brains could become more complexly wired and remain more plastic than the brains of less mercurial sorts. That heightened connectedness between one neural neighborhood and the next, as well as a continuing receptivity to new information, new sensations, may allow a person with a mood disorder to affiliate seemingly incongruous ideas and to reimagine the ordinary into the extraordinary — the essence of artistic creation. People who have experienced emotional extremes, who have been forced to confront an enormous range of feelings, and who have successfully coped with those adversities could gain a richer organization of memory, a richer mental palette. Moreover, the excessive energy of a manic episode may give birth to a volcano of ideas that the mind can then shape into something meaningful during the less frenetic, more skeptical moments of a depression or spell of normality.

Added evidence that manic-depression is not entirely a scourge and may, at least in milder forms, bring benefits to those who bear it, is the prevalence of the condition. It cannot be explained by chance alone. Family and twin studies suggest that the disorder is inherited, though whether it arises from a mutation in a single gene or a handful of genes nobody can yet say. What is known is that if manic-depression were a random, erroneous, and thoroughly deleterious genetic disease, you would expect to see it in only one in three thousand people. Instead, the incidence is at least one in a hundred or higher, whether you look at people in New York City or the Kalahari Desert, indicating that the condition is here for a reason. From an evolutionary angle, then, manic-depression might better be considered a trait than a disease, a genetic variation on a temperamental theme that in prehistoric times conferred strong advantages on those who inherited it. One can only guess at the nature of such advantages, but surely ancestral humans who were unusually creative, energetic, adroit at solving problems, fearless in the face of the unknown, and at times grandiose and self-promotional would have had one meaty, pelt-clad leg up on their mild contemporaries. Even today, manic-depression is associated with the trappings of success, for the disorder is somewhat more prevalent among those of high socioeconomic standing than those in the lower tiers of the pyramid.

How manic-depression and other mood disorders are reflected in the topology of the brain is only now coming to light. Preliminary neural imaging studies indicate that different regions of the brain are perturbed during either manic or depressive episodes, bolstering the idea that a bipolar mood disorder is a global arouser of mental activity. In one experiment, volunteers were given intravenous doses of procane, a drug that can elicit emotional responses from euphoria and a speedy sort of energy to anxiety and depression. Using positron emission, or PET, scans to measure relative blood flow in the brain, the neurobiologists found that when the volunteers reported feeling depressed from the procane, certain regions of the brain's limbic system showed diminished activity, among them the amygdala, the orbitofrontal cortex, and the cingulate gyrus. These are the structures of the brain that control such emotions and behaviors as anger, pleasure, and aggressiveness.

By comparison, when the drug induced in the volunteers a euphoric state similar to mania, limbic activity increased, as did the responsiveness in structures of the midbrain that interact with the limbic system, notably the hypothalamus, master regulator of sexuality and other psycho-bodily states. That a stimulated limbic system and hypothalamus might yield greater associative and imaginative powers is not surprising. The entire neural region is a major switching station, turning external stimuli into emotional responses, and emotional responses into action. It's also the part of the brain that evolved parallel with mammalian sociability, encouraging individuals in a group to recognize and respond to one another. As the region that helps us take in the new, integrate it with the familiar, and generate from the exercise a novel response, the limbic system is, in a sense, a microcosmic creator.

The portrait of the mood-disordered brain encompasses more than quirks in the limbic system. Other brain-imaging studies of manic-depressives have shown distinctive patterns in the metabolism of the prefrontal cortex, the most advanced part of the brain and the seat of human intellect, indicating that changes in the biological underpinnings of thought occur parallel with shifting emotional and physical states.

In imaging studies of patients with brain lesions that leave them either constantly weeping or constantly laughing, neurologists have noticed hemispheric distinctions. Those who cry without ceasing have damage to the left hemisphere of the brain, considered the half in command of language and rational thought;

those who are uncontrolled laughers display lesions in the right side of the brain, where nonverbal information is processed.

If in manic-depression or its gentler variations artists list first starboard and then portside on the blinding swells of the imagination, small wonder that, on returning, they distill paradox into beauty.

34

SCIENTIST AT WORK:

VICTORIA ELIZABETH FOE

VICTORIA ELIZABETH FOE seems to have the wrong voice for the rest of her. On the telephone she sounds small and timid, like a character from a moody Anita Brookner novel who sits by herself at a corner table, sipping tea and dispassionately reviewing the minor disappointments of her life. But in person, when Dr. Foe loosens up and starts talking about her work as a developmental biologist who spends days upon nights peering into a microscope, observing the earliest moments of embryonic life, and charting every throb and tremble of every individual cell, she expands to epic dimensions and lets her superlatives fly.

"There's a deliciousness and a delight to looking at embryos," she said. "It's a celebratory act, an act of enormous pleasure." Later, she said: "It's like a diver going down into the sea. You notice something new and something totally amazing every single time. It's just a wonderland." Still later she said: "This is biology's golden age. It's analogous to cathedral building of a thousand years ago. We are building and building this great edifice. Some of us are building arches, some are painting murals, some are carving in stone. I feel enormously privileged to be alive now and to be a part of it."

The woman herself has so much flair and presence that everything else looks stunted by comparison. She wears a long peasant skirt, a peasant blouse, and cowboy boots that surely lift her up to six feet. She has a wild spray of long black hair and features that are at once delicate and strong.

Dr. Foe, a research scientist at the University of Washington, is celebrated among developmental biologists and drosophila geneticists, who study the fruit fly for perceptions about how all animals grow. Her peers know her work and cite it in papers of their own. They know she is the one who made the seminal observation that, early in development, different groups of cells in different regions of the formless embryo begin dividing at markedly different rates, a discovery that casts light on how a monotonous bleb of tissue gives rise to the complexity of body and brain. She is so highly regarded that the National Institutes of Health gives her a supporting grant that is independent of any affiliation with a university or institution, the only such grant it awards. Dr. Foe has also won a five-year MacArthur Award, the so-called genius prize that comes with no strings beyond the expectation of continued brilliance.

She is a woman who has forged a new path for herself in science. Most other biologists either work in academia, trudging down the track to tenure and gathering ever more graduate students, postdoctoral fellows, and technicians along the way, or they take jobs in industry for steady salaries.

Dr. Foe has done neither. She has never sought a professorship, because she doesn't want her life to be consumed by administrative duties that would take her away from her research. She has never wanted to work for a company, out of a dislike of being bossed. Nor does she like playing boss. Until she turned forty-eight, in 1993, and decided to take on one graduate student, she'd never had a technician, student, or anybody else work for her. Her grant from the NIH supplies her salary and money for equipment and biological materials, which means that her arrangement with the University of Washington is relatively loose. In essence she is a self-employed scientist, and what she has done she has done largely on her own, as Foe Incorporated, a one-woman band. Her style is fundamentally different from the way most science is done today, closer to that of Gregor Mendel, the solitary pea shuffler, than to the large, disciplined teamsmanship that characterizes such contemporary enterprises as that international goliath, the Human Genome Project.

Yet in her tremulous telephone voice lies a telling clue to her character. Despite her accomplishments, the courage of her unorthodoxy, and her conviction that the work she is doing is grand and profound, Dr. Foe has all the ego strength of a teenage girl at her first sock hop. She gets nervous easily and sweats readily. She is always afraid that her grant will run out and not be renewed. She works maniacal hours and then worries that she has not done enough, or is wrong or obtuse or muddled in her thinking. She considers herself a dull and inept speaker, although everybody who hears her talk is enraptured. "She absolutely knows what she is doing is the right thing to do," said Dr. Garrett M. Odell, a developmental biologist at the University of Washington. "But that is coupled with this amazingly persistent insecurity about whether she is good enough to be talking to smart people. Even now, she's worried that the MacArthur people are going to call her back and say, 'Are you VICTORIA Foe? Oh, sorry, we meant VICTOR.'" Dr. Odell is Dr. Foe's collaborator and her live-in mate, although that largely means living in at the lab.

The mix of fear and fortitude is only one of many contradictions that Dr. Foe embodies. In an era when most basic biologists take a reductionist approach to their craft, breaking down a problem into its smallest possible components, Dr. Foe is like an old-fashioned naturalist who observes the nuances of the entire organism as it goes about the business of growing. She looks at embryos of fruit flies, blowflies, mosquitoes, frogs, hornworms, fish; dusk cedes to dark, the dark to dawn, and still she keeps looking. Yet far from being an imprecise empiricist, she swiftly adopts new technologies and uses molecular biology to make her observational work ever keener.

"I use twentieth-century techniques to carry out a nineteenth-century approach, where I spend hours just looking," she said. The technology is now taking on a twenty-first-century flavor, as she and Dr. Odell hook up her observational microscope to computers and controls that will allow Dr. Foe to mark individual cells of a fly embryo and follow them through the entire odyssey from their emergence in the blastula to their destiny as constituents of antenna, foot, thorax, or eye. The enterprise is expected to be completed by the latter half of the 1990s, and the cellular atlas that results will be the most detailed record ever produced of embryonic growth, far outstripping in complexity the famed fate map that scientists have made of the much more primitive laboratory roundworm. Dr. Foe, as well as the many other fruit-fly geneticists who await the production of her master atlas, will then be able to use the blow-by-blow descriptions of cell fate to determine which genes control the behavior of cells as they mature. And for scientists, it is an article of faith that finding the genes behind the growth of fruit-fly larvae will hasten the discovery of the equivalent developmental genes in humans.

The observational approach also suits Dr. Foe's aesthetic sensibility. "There are always roads you didn't take, other paths that beckon, and for me the other road was art," Dr. Foe said. "If you look at my science, it's extremely visual." She makes detailed drawings of what she sees, and her finely delineated and brilliantly colored diagrams of fruit-fly embryos are so exacting and sensual that it is a pity they are confined to scientific journals. Dr. Foe is also a perfectionist. While other scientists rush to put out as many publications in as brief a period as possible, she will work for years on a project, not reporting her results until she is satisfied that the findings are solid and complete. As a result, her curriculum vitae is shorter than that of many scientists her age, but most of the papers listed are more like books than reports. In publishing one of her exhaustive papers, the journal *Development* warned other would-be authors not to take similar liberties in length unless they were prepared to match her rigor and breadth of research. "Her approach takes the sort of patience that almost nobody has anymore," said Dr. Odell. "She's monastic."

Yet here is another contradiction. Dr. Foe is a scientific loner, but she has a strong political conscience. One reason she never sought a tenure-track position in science was her wish for the freedom to switch gears quickly when she wanted to pursue political activities. As a graduate student at the University of Texas at Austin, she dropped out for a year and a half to work as a political aide and help overturn the state's antiabortion law. She was involved in the women's movement and the anti–Vietnam War movement. More recently, she and Dr. Odell rallied vigorously against the Persian Gulf war, and went to Canada, where Dr. Foe has a plot of land, to protest the government's plan to permit logging in an old-growth forest. "Most scientists try to ignore the rest of the world, but she has a tremendous guilt complex," her mentor, Bruce Alberts, told me. "She feels guilty about doing research all the time and allowing people to starve or go to war." Even in her science she finds a political moral. "The wonderful lesson to come out of biology in the last few years is that the same genes, the same parts, turn up again and again, from one species to another," she said. "The important point to realize is that we're all made of the same fabric, we're part of the same web, and there is some humility in the idea that is appropriate."

Dr. Foe came by her capacious perspective in childhood, as her family moved from Wyoming to Mexico to England and back to the United States, her father changing careers from law to farming to teaching. Her father was a shining knight in her life, fascinated by everything, and stimulating the same intellectual hunger in his three children. He died of congestive heart failure when she was twenty-one. Dr. Foe still misses him and feels deep sorrow that he did not live long enough to see what she has discovered as a diver into microscopic realms. "He put up with our larval states and then didn't get to enjoy us as human beings."

Dr. Foe's unconventional scientific career can be traced to her marriage to Dr. Michael Dennis, a neurophysiologist at the University of California. When her husband decided to quit science and become a sculptor, Dr. Foe, then a postdoctoral student, chose to leave California with him and his daughter and move up to Denman Island in Canada. Horrified that Dr. Foe, too, might leave science, her adviser, Dr. Alberts, helped her arrange to work out of the Friday Harbor Laboratories in Washington, a spectacularly beautiful research station not far from the Canadian island. "I thought it would be an incredible waste if she went up there and did pottery or something," he said. Dr. Foe completed her postdoctoral training at Friday Harbor, and then, in the mid-1980s, persuaded the NIH to give her the independent grant she still receives today. Dr. Foe and Dr. Dennis have since divorced, although she continues to see him and his daughter and has built a one-room cabin for herself on Denman Island.

Her talents have not been squandered. In the findings that ensured her reputation, she discovered in the late 1980s that early in the development of an embryo, a spectacular transition occurs. All the cells of the embryo grow in unison through the first thirteen rounds of division, splitting in half and in half again as a single pulsating bundle, but the synchrony breaks down at the fourteenth division. At that point, different groups of cells begin dividing at markedly different rates. These various arenas of division are called mitotic domains (after mitosis, or cell division), and there are twenty-five of them in the early fly embryo, with clear boundaries between one domain and the next. These domains are previews of coming attractions, the first visible signs of cell specialization that eventually will yield distinct organs. Much farther down the line in embryonic growth, one domain will give rise to the nervous system, and another to the creature's limbs. "It looks as if the embryo has painted on it its own fate," Dr. Foe said. Yet even with this window on cell determination, scientists cannot tell exactly how the domains give rise to body components. "We have a rough fate map, but we don't know any of the little territories in detail," she said.

"We may know this domain is in an area that gives rise to the nervous system, but we don't have a clue to which component of the nervous system it makes."

Gathering those details will consume the next phase of her career as she traces the lives of individual cells within the different mitotic domains — monastic, often alone, but delighting in the spectacle unfolding before her.

35

SCIENTIST AT WORK:

MARY-CLAIRE KING

MARY-CLAIRE KING, a geneticist of international renown and neutronic stores of energy, is sitting in her sunny, unprepossessing office at the University of California in Berkeley, talking about the current monarch of her many passions. She is trying to find the gene for hereditary breast cancer, a gene that could be of great significance to hundreds of thousands of women who are at risk for early onset of the disease. She has been seeking the gene for seventeen years, weathering the skepticism of her colleagues and often her own doubts as well. In 1990, she found the approximate location of the gene, and now she and her students are struggling to home in on the trophy proper. She wants it very, very badly, and she believes her laboratory is very, very close. She also knows that other labs have since joined the race, and she would dislike seeing some newcomer step in at the final hour and seize victory.

"It could be there right now, sitting on one of our plates," she says, referring to the petri dishes where segments of isolated genetic material await analysis. Her voice intensifies, and her deep dimples disappear along with her smile. "We're obsessed with finding that gene. I want it to happen in our lab."

One of her students, a young Asian, pokes his head in the door, grinning broadly, and says he has something to tell her.

She excuses herself and joins him in the room next door. Suddenly, a crow of delight fills the halls: "Yes! Oh, yes! That's WONDERFUL!" She returns to the office, her face glowing, and I wonder for a moment whether I've won the science writer's sweepstakes: being on site for a spectacular discovery. Has the gene been found? I ask. Are the scientists even nearer their goal than she suspected? "He just told me he's getting married," she says. "I am so, so happy for him."

That Dr. King should react with untethered joy to her student's joy is hardly surprising. Though she was trained as a mathematician and is now a molecular geneticist as committed as any basic researcher to rigor, abstraction, and competition, nearly everything she has ever chosen to work on has reflected, at its core, her deep sense of humanity. She first earned fame by working in Argentina with a human rights group, the Grandmothers of Plaza de Mayo, attempting to reunite with their families those children who had been kidnaped in the 1970s and early 1980s by the military junta. By analyzing genetic material from the children and comparing it with the genes of grandmothers and other relatives who survived Argentina's eight-year "dirty war," Dr. King and her co-workers were able to prove that many children had been snatched away as infants and given to other families when their real parents were either shot outright or had mysteriously disappeared.

Dr. King also has immersed herself in the case of El Mozote, a village in El Salvador where, in 1981, at least 794 peasants, many of them children, were massacred by American-trained Salvadoran soldiers. The first skeletons of the victims were dug up in 1992, and the government of El Salvador has agreed to permit a thorough forensic analysis of the remains once the exhumations are complete. Dr. King and other researchers are trying to identify the skeletons by comparing DNA extracted from bone and teeth with that of living relatives, for possible use in criminal proceedings. "This case is much more difficult than the Argentinian one," she said, "because there are very few survivors of the massacre," and thus nothing to compare the DNA to.

Dr. King, who was born in 1946, is an unreclaimed liberal, and she is delighted that her headquarters, wedged in the middle of a building devoted to forestry science, has a bit of history to it. In 1970 in that very room, as a graduate student at Berkeley, she and other students organized a letter-writing campaign to protest the American invasion of Cambodia: they gathered thirty thousand signatures from voters in Northern California. Yet Dr. King is not given to political posturing, and she is amused to say that she is now collaborating with the United States Army. "We're working with our own government, I pale to say, on MIA cases," she said, including an attempt to identify a man, shot down in a fighter plane during World War II, whose body was preserved in the bog where it landed. The King lab does not consider itself a forensics lab proper, but its researchers have perfected a means for extracting DNA from teeth, taking it from the nerve pulp that remains. And teeth, it turns out, are better preservers of genetic material than are bones.

Dr. King is also enough of a pragmatist to have hoisted herself up to the summit of mainstream science. She was a strong candidate to replace Dr. James Watson as the director of the Human Genome Project, the celebrated enterprise to map and analyze all 100,000 human genes; the job went instead to Dr. Francis Collins, a geneticist at the University of Michigan, who is collaborating with Dr. King on the quest for the breast cancer gene. Dr. King was asked to apply for the job as head of the National Institutes of Health to replace the departing Bernadine Healy, but declined to be considered. "I'm not interested in a job with that level of administrative responsibility," she said. "It would be too far removed from what I love to do, which is science."

Yet as pure scientists go, Dr. King has a pronounced bedside bent. She and two other researchers published a report in *The Journal of the American Medical Association,* anticipating the isolation of the gene behind early-onset breast cancer, and discussing the possible options for those who carry the mutant gene, an estimated 600,000 women in the United States. Such women are at extremely high risk of contracting breast cancer before the age of fifty, and must think carefully about whether to take such drastic measures as having their breasts removed prophylactically or enrolling in the current trial of tamoxifen. Scientists hope the drug will help prevent many breast cancers, but its effectiveness is unknown, it has several potential health risks, and it will throw a young woman into early menopause.

Dr. King's lab is also doing two projects on AIDS research, trying to ascertain whether genetic variations could explain why some people survive with the disease much longer than others. Her team is studying the genetics of systemic lupus, an autoimmune disease in which the skin and joints are progressively destroyed, and it is hunting for the gene behind hereditary deafness.

Dr. King is a vigorous proponent of the Human Genome Diversity Project, spearheaded by Dr. Luca Cavalli-Sforza, a population geneticist at Stanford University. The researchers plan to sample genetic material from some four hundred human populations worldwide, with an emphasis on the oldest and least intermixed people, like the Basques of Spain and the Ket and Gilyak of Siberia. By scrutinizing the chemical runes of genes, the researchers hope to answer questions of evolutionary, linguistic, and anthropological sweep: Where did modern humans come from? How did they migrate across the globe? Did genetic changes in any way correlate with language variations? And can genetic discrepancies explain differing rates of disease in different countries? Dr. King spends several days each month in Washington, in part to lobby for money to support the enormously complex effort.

This harlequin collection of projects is conducted by a relatively small lab of twenty people, including Dr. King. She also teaches graduate and undergraduate courses, including a freshman genetics class for nonscience majors. And, rare for a research professor, she sees teaching not as drudgery but as pleasure. Yet, in spite of it all, she manages to look young for her age, with a slab of dark hair that seems almost immobilized by its own thickness. Doesn't she ever feel overwhelmed? "Of *course* I feel overwhelmed!" she says, her voice rising up in a whoop reminiscent of Julia Child's. "What does being overwhelmed have to do with it?" She walks and talks so swiftly that by comparison one feels trapped in resin.

Dr. King traces her scientific style to her mentor and thesis adviser at Berkeley, Dr. Allan C. Wilson, an intellectual firebrand who died of cancer in 1991 at the age of fifty-seven. Dr. Wilson was famed for his work on the so-called genetic Eve, a woman who supposedly lived about 100,000 years ago in Africa and is the theoretical mother of all humans alive today. Those who worked in Dr. Wilson's lab mastered the art of attacking evolutionary puzzles with molecular artillery, relying on the genes sequestered in the mitochondria, the tiny powerhouses of the cell. Dr. Wilson also stopped Dr. King from quitting science almost before she got started. "I could never get any of my projects to work, and I was very depressed and distracted," she said. "He said, if everybody who couldn't get anything to work dropped out of science, there would be no science." Thus reassured, she completed her Ph.D., showing, to the shock of herself and the entire scientific community, that humans and chimpanzees have more than 99 percent of their DNA in common. "I kept thinking I had a negative result, because I wasn't finding any molecular differences," she said. "Then I realized that's because there were almost no differences."

From there she went to Chile with her zoologist-husband, Robert Colwell, to teach, but they returned to the United States after the leftist government of Salvador Allende was overthrown. As a result of her experience in South America, her familiarity with the language and people, when the *abuelas* of Plaza de Mayo sought the help of scientists to solve the problem of the missing children, Dr. King became the molecular geneticist on the case. The work was alternately grueling and inspiring, demanding frequent trips to Argentina, eighteen-hour days, and the spine to stand up against the surly and grudging military there. The Argentinian project continues, and likely will continue into the foreseeable future; so far, fifty-three children have been reunited with their original families, but another 150 have yet to be found. They would now be young adults, and could be anywhere: in Latin America, across Europe, in the United States. Throughout the project, Dr. King was spurred by the knowledge that the kidnaped children were the age of her own daughter, Emily Colwell. Dr. King's marriage broke up when Emily was five, leaving Dr. King a young, single mother struggling to succeed in a field known for its lengthy, irregular hours and its pitiless pace. She believes that one reason her marriage failed is that, as a scientist, "you can only do two things out of three. I was a young mother, a young scientist, and a young wife. Something had to collapse, and it was the marriage."

In her own lab, Dr. King tries as much as possible to accommodate those with families. True to the capacious spirit of Berkeley, she has always had many female graduate students and postdoctoral fellows in her lab, as well as African-Americans, Chinese, Latinos, gays. "I was in this business for years and years before I had a straight white male graduate student," she said. Yet there is a limit to how freewheeling any high-status lab can be, and many of her students practically bunk down benchside, especially those involved in the breast cancer project, where outside competition is sharpest. Dr. King says she hates competition, and she sees it as one way in which men have put a needlessly and tediously masculine stamp on the profession. She once finished a talk at a scientific meeting by pulling out a pencil case shaped like a shark, given to her by the child of a friend. She placed the case on the table and said, "This is for all the sharks in the audience," a line that earned her scant laughter and less applause. But as the men who know her are quick to point out, Dr. King is no wilting lily. She speaks her mind, goes after what she wants, and does not cede ground. "When it comes to competition," says Ray White, a friendly adversary in genetics, "Mary-Claire is at no disadvantage, no disadvantage at all."

Postscript. In the fall of 1994, the breast cancer gene was finally found, but not by the King lab. The victor of the race was Mark Skolnick, of the University of Utah, and a team of forty-four co-workers. Although Dr. Skolnick and Dr. King were known to dislike each other — the result of a collaboration that had collapsed in bitterness twenty years earlier — Dr. King graciously complimented him on his good work in all the interviews she gave, and said that, while she had expected to feel bad about losing, when she actually lost she felt nothing but relief that the race was over. Now geneticists face a much stiffer challenge: to transform the discovery of the gene into something of meaning for women everywhere who fear for their breasts and their lives.

36

AT THE SCIENCE MUSEUM
WITH STEPHEN JAY GOULD

STEPHEN JAY GOULD is in the cafeteria of the California Academy of Sciences, a snug West Coast rendition of the American Museum of Natural History in New York, and he is about to try what generations of children have delighted in doing whenever they are in a lunchroom: blow the wrapper from a drinking straw at the ceiling and make it stick. He dunks one end of a straw and its wrapper into a glass of water, and the paper forms a silly, soggy proboscis at the tip — the sticky part. He carefully tears the other end of the wrapper so that he can hold and blow at the same time. Oh, yes, he loves the straw trick, this esteemed Harvard University professor of geology, biology, and the history of science, this tireless writer of densely elegant science essays that reach a huge popular audience, and, it so happens, this international authority on a small tropical snail called Cerion.

But when he puts the straw to his lips and gives it a puff — *Phht!* — nothing happens. The wrapper is too riddled with tiny holes to hold the air needed for propulsion, and Dr. Gould tosses the whole thing aside in cheery disgust. He has thought long and deeply about evolution, and he dislikes the widespread misconception that evolution equals progress and inevitably leads all creatures toward a more perfect state of being; but he knows devolution when he sees it, and this product is it. First, they substituted plastic straws for paper ones, he complained. "Now, they can't even make a decent wrapper. It's outrageous." Shrugging, he picks up a plastic spoon and goes after his soup.

Dr. Gould is in San Francisco promoting *Eight Little Piggies,* his sixth collection of essays. Most of the essays first appeared in *Natural History,* the magazine that he has written for every single month without a break for the last nineteen years, as he proudly announced in one of those very essays. He writes for other periodicals as well, including *The New York Review of Books, Discover,* and the British journal *Nature.* He has also written books on single topics, among them *This Wonderful Life,* a best-seller about the reinterpretation of the Burgess Shale, a group of rocks from a quarry in western Canada that preserves an enormous diversity of fossils from about 550 million years ago. He types and types: he is the Anthony Trollope, the Joyce Carol Oates of science writing. And he teaches: his classes at Harvard are often standing room only. And he does research and writes scientific papers. Even an attack ten years ago of mesothelioma, an often fatal cancer of the stomach and pelvic cavity, did not slow Dr. Gould down appreciably. He says that now that he has long since conquered the cancer, he wants to finish the opus he began before the disease struck: a stringent reconsideration of Charles Darwin's theory of evolution. He's always been like this, friends say. He works, works, works.

By scientific standards, Dr. Gould is a celebrity, recognized by passersby in the academy cafeteria, who come over to the table, pump his hand, and bellow, "Professor Gould! What brings you to California?" He was a pioneer among working scientists who feel it their bounden duty to share the bliss of science with the untutored masses. He has not yet had the monumental publishing success of Stephen Hawking, the astrophysicist whose slim volume *A Brief History of Time* sold 1.775 million copies in the United States and Canada alone and spent years on best-seller lists here and abroad. Dr. Gould has not yet been the host of his own television series, as Carl Sagan, the Cornell astronomer, was with "Cosmos." And, though his style is accessible, sometimes charming, and invariably free of condescension, it lacks the radiance and music of the work of the late Lewis Thomas, the physician and author of *The Lives of a Cell*. Nevertheless, the Stephen Jay Gould fan club is international in scope. His books have been translated into fifteen languages, among them French, Japanese, Hungarian, Finnish, and Greek. His writings, more than anybody else's, have transformed Darwin into a household icon, a scientific saint on a par with Albert Einstein.

Dr. Gould loves writing, but he dislikes book tours, which is why he is feeling a bit fidgety and peevish during his visit to the academy, part of the circuit. A fan, a young man of about twenty-five, comes over to the table and introduces himself, and when he departs, Dr. Gould says, "What is it about this younger generation? Whenever they say something, the voice rises at the end in a kind of question. Everything they say sounds as if they're asking for confirmation." He pretends to be holding a telephone receiver to his ear. "It's like this: 'Hello? Dr. Gould? This is Amy? Of the AP?' Well, yes, I didn't doubt it for a moment."

Stopping by the dinosaur exhibit at the museum, Dr. Gould waves his hand dismissively and says, "Dinosaurs have become boring. They're a cliché. They're overexposed." As he watches a group of children dash raucously from one hands-on display to another, he mutters, "Are they learning anything, or just pushing buttons?" At the hanging pendulum, a giant steel ball suspended from a wire in the ceiling, he says: "I've never understood why every science museum in the country feels compelled to have one of these. I still don't understand how they work, and I don't think most visitors to the museum do either." The pendulum is supposed to show that the suspended ball keeps swinging in a straight line while the earth rotates beneath it, but Dr. Gould points out that the pendulum itself is attached to a building that is rotating with the earth, so why should the axis of the ball not be rotating as well? An excellent question, which the display placard fails to answer.

Yet even in his restive vein he is a nimble raconteur, who talks the way his essays read. He picks up a filament of an idea, follows it a short distance, loops it together with another insight and yet another, until enough strands have been threaded in to make a plushly coherent pattern. "Everybody has some curious little mental skill," he says. "Mine just happens to be making these connections. If you're lucky, you learn to convert that skill into a professional advantage. Otherwise, it's just a party trick."

As a prominent critic of overexuberant genetic determinism and the attempts to judge innate abilities and intelligence through standardized testing, Dr. Gould is tired of continuing efforts to resolve the old nature-versus-nurture debate. He thought he had addressed these issues once and for all in his 1981 best seller, *The Mismeasure of Man,* but everywhere he goes people ask him: How much of our intelligence is inherited? How much of it is the result of education? Is criminal behavior innate or learned? We want numbers! We want answers! Dr. Gould emphasizes that biology and environment are inextricably linked, that the influence of one cannot be disengaged from the influence of the other. "It's logically, mathematically, philosophically impossible to pull them apart," he says. "It is a true union of influence, but I despair of getting people to understand that. It's unfortunate that there's a linguistic similarity between the words *nature* and *nurture.* That's helped keep this ill-formulated and misguided debate alive."

Dr. Gould's ideas on evolution and the design of nature, while broadly influential, have nettled a few colleagues. This is true of the so-called punctuated-equilibrium theory — the notion he and Dr. Niles Eldredge have proposed that evolution proceeds in ragged fits and starts rather than unfolding smoothly and gradually. In a scientific world that portrays the individual organism as an island unto its wretched self, pitted for survival not only against predators that may eat it or parasites that may deplete it, but against all others of its kind, Dr. Gould has had the chutzpah to suggest that perhaps some attributes we see in nature evolved for more communal purposes, to benefit not merely the individual, but an entire species. He has argued, in other words, that the old truism "for the good of the species," an idea scornfully rejected by many current evolutionary biologists, may have a certain validity to it after all.

Some of his detractors grumble that Dr. Gould has not really contributed that much to research; some say that what he has contributed has often been bad; others complain that his essays are a touch self-serving. Niles Eldredge believes that Dr. Gould is a victim of hype from two different directions, exalted as the oracle of evolutionary theory, scalded as a tired demi-socialist and rank popularizer. "Some people would like to dismiss Steve," he said. "They're always trying to come to terms with him, to emotionally metabolize him."

For his part, Dr. Gould says he doesn't dwell much on his detractors, and he insists he is respected by the great majority of his colleagues. He still has a tincture of New York scrappiness and defensiveness, the boy from a lower-middle-class neighborhood in Queens whose father, a court stenographer, was a self-taught man and never stopped feeling inadequate because he lacked a college degree. Dr. Gould lives in Cambridge, Massachusetts, with his wife and two children, whom he rarely discusses in public. Nor will he talk about his private life in general, although he has made widely known his passions for baseball, Wagner, and Mozart.

Mostly, though, Dr. Gould seems comfortable: pudgy in a comfortable sort of way and a lover of comfortable foods like french fries and chili. He says he has a new appreciation of the imprecision, the slop, and the redundancies that everywhere can be seen in nature. *Eight Little Piggies* plays on this theme of repetition, slack, and fortuitousness in the natural world. The title essay, for example, considers why modern vertebrates have five digits on each limb, rather than eight like some prehistoric animals, and concludes that it was an accident of evolution. And he writes that the excess genetic material we carry around in our cells, material that many have dismissed as junk DNA, may be the stuff of evolutionary ingenuity and change.

"In my youth, I was very much into this macho idea of science as rigid, hard, quantifiable," he says. "Now I'm interested in the beautiful and quirky contingencies of nature." Nothing stays rigid forever, of course. The straw wrappers that he blew onto the cafeteria ceiling of the Natural History museum in New York more than four decades ago dried into little stalactites and stayed in place for years, but when he looked up as an adult, he said, not a single paper fossil of his boyhood could be found.

VII

DYING

37

CELL DEATH

AS THE KEY TO LIFE

IT IS A PARADOX that neither asks nor accepts resolution: the indispensability of death to the perpetuation of life. Whether we are growing with the grace of youth or plodding on with the fatalism of age, our cells die daily by the millions — and for that we must be grateful. As a baby's brain develops in the womb, about 80 percent of the nerve cells perish within hours of their creation, their lacy tendrils of no use and possible harm to the final organ. The vestigial webbing that forms between a fetus's fingers must be dissolved before birth. In adults, immune cells that would mistakenly assault the body's own tissue must be extinguished without a fuss.

Cell death is universal to life, yet science has long neglected the problem as either tedious or overwhelming. A dead cell is almost impossible to study — it is, after all, a lifeless thing, inactive, post-interesting. Yet a dying cell was viewed as *too* interesting; it was a forest fire, in which any single flaming tree is lost amid the blinding mayhem, beyond reduction and analysis.

Recently, though, biologists identified a type of cell death that proves amenable to study, a species of passing known as programmed cell death, or apoptosis. Programmed mortality turns out to control the elimination of cells and tissue in a broad variety of circumstances, and the latest revelations have lent new life to the study of death. Biologists have found genes that work like time bombs within undesirable blood cells, setting off chain reactions that quickly degrade the immune cells before they can expand into hazardous colonies of identical cells. They have detected proteins that trim back the abnormal bulges of the body's liver and prostate, that shuck off a caterpillar's crawling muscles when the insect is ready to emerge as a moth, and that help whittle away female genitalia from an animal destined to be a male.

These discoveries could yield fresh approaches to the treatment of pathological cell death, like the massive brain degradation found in people with Alzheimer's disease, Parkinson's disease, Lou Gehrig's disease, and other degenerative disorders. By understanding the genes that cause cell death, we may also gain some clue to what happens when those genes fail and the cell assumes the immortal mantle of cancer. A precise understanding of the death of a single cell may elucidate why the whole person — you, me, all we see — eventually must die.

To their surprise, researchers have found that in many cases the death of a cell is not a chaotic conflagration, as had long been believed, but a highly choreographed genetic program, much like the one that provokes an immature cell to become a working member of the pancreas or the kidney. It is a great challenge for a cell to expire swiftly and with style, and a cell scheduled for destruction must first activate a dozen or more genes to perform the deed. It's as though the cell tells itself, I'm going to die, and in a deliberate, focused manner carries out the decision. This does not mean that the act of cell death is in any sense sedate. Quite the contrary: a cell in apoptosis departs with an inspiring bang, blowing open and exploding its contents into the bloodstream for rapid absorption.

While scientists concur that the subject of apoptosis is very much alive, they have yet to agree on how to pronounce the term. Some start the word with a long *a*, others with a short; some pronounce both *p*'s, others drop the second one. The term was coined by Dr. Andrew Wylie, of Edinburgh University in Scotland, who first described the features of this type of cell death and named the process after the classical Greek for "falling from," as leaves fall from trees. But biologists now joke that the word should really refer to hair falling from the scalp, since the male scientists in the apoptosis field all seem to be balding.

The discovery of apoptosis overturns a longstanding notion that cell death is an easy thing, an absence of life, a cessation of input — the cell's default mode. And it remains true that, in some cases, like the tissue degeneration that follows extreme trauma to the body, cells do simply swell up and die, randomly and sloppily and without any apparent program. Immunologists, however, were bothered by this explanation for the death of the cells they cared about — the T and B cells of the immune system. The body propagates these cells by the quartload on a regular basis, each fashioned to assault a foreign protein of a slightly different shape. But the vast majority of the immune cells, perhaps 95 to 98 percent of them, turn out to be bad seeds, with a shape that would attack the body's own organs or do something else unsavory to the home plant if they were allowed to survive. Immunologists knew the body got rid of these cells as quickly as it made them, and they began seeking the murderers — the body's hit squads, presumed to carry out the business of eliminating undesirable immune cells.

In searching for the murder weapon, scientists were led instead to a suicide note. The bad T and B cells weren't getting killed off; they were killing themselves. Soon, the stanzas of the cellular swan song had been roughly jotted down. Heeding a signal that has yet to be identified, a cell with potential to do the body harm stoutly initiates its apoptotic program. It turns on enzymes within the nucleus that act as pruning shears, snipping the chromosomes into tiny uniform pieces; it activates other enzymes to rip holes in the cell's protective membrane; it sends forth a keening chemical cry that attracts the body's cleanup crew of macrophages to the site; and, finally, it ruptures itself apart and into the maws of the macrophages. As one biologist phrased it, apoptosis is a ritualistic dance of death.

Grim though it sounds, the lack of apoptosis is grimmer still. Scientists have seen that in some types of human lymphoma, a mutation in a gene lying along the apoptotic pathway is to blame. The gene, called bcl-2, operates under normal circumstances to block cell death in some of the body's B cells, which can then serve as part of the immune system's long-term memory — its ability to recognize and attack repeat microbial offenders. However, when the gene is mutated, it remains permanently active in all B cells, not just some of them; as a result, white blood cells don't heed the command to kill themselves when they should, but mill around indefinitely and eventually gather to cancerous proportions.

Simpler organisms than humans also rely on apoptosis. In the translucent, millimeter-long roundworm, *C. elegans,* exactly 131 cells out of its total body count of 1090 cells are born only to die. Biologists have found a number of death genes that carry on in those 131 cells, and they've placed the genes in a sort of demolition hierarchy, observing that some refuse to begin the fatal games before other genes have commanded them to do so. Quoting "Alice's Restaurant," the scientists say the first step is kill, kill; the second is to get rid of the body; and the third to clean up the bloody mess left behind. So different genes come into play at each stage, some breaking the cell apart, and others encouraging neighboring cells to consume the debris.

Why does the worm's body bother making the cells in the first place, if they are only going to sacrifice themselves? The cells, it seems, are like Michelangelo's marble block, from which the glorious wormly form will be sculpted. Much of the excess tissue arises in the worm's genital bud, and the cells will die off in one way or another depending on whether the nub is to become a he-bud or a she-and-he-bud, a hermaphrodite.

In another case of death preceding metamorphosis, the tobacco hawk moth uses sheets of giant muscle cells to emerge from its cocoon. Once the moth has pulled itself free of the incubating cuticle, it no longer needs the muscle tissue, and the cells degrade overnight, apoptosis-style. When they dissected the cells, which, at five millimeters in diameter, are easily visible to the naked eye, scientists found several proteins that switch on to dramatically high levels just before cell death begins. One such activated executioner is the protein ubiquitin, which fastens on to hundreds of other proteins within the cells and tags them for degradation. With so many of their proteins destroyed, the cells themselves rapidly die. The correlation between ubiquitin and mass cell suicide may not be limited to hawk moths. Traces of ubiquitin litter the dense neurofibrillary tangles found in the brains of Alzheimer's patients, suggesting that the protein was stimulated to bring about the untimely demise of nerve cells. Why the protein was rallied to its murderous call, though, and what can be done to prevent its action, remain questions on the scientists' lengthy docket. The mysteries of cell death won't give themselves up without a fight.

38

MYC GENE, ARBITER OF DEATH
OR LIFE

AMID THE TWISTED WRECKAGE of chromosomes found at the heart of nearly all human cancer cells lies a handful of molecular aberrations that seem to be, not the incidental debris of malignant transformation, but the fundamental defects that spawned the cancer in the first place. Among the most dangerous and widespread mutations yet found are those which disrupt a gene with the folksy name of myc, pronounced mick. Whether in tumors of the breast, brain, bladder, blood, lung, colon, or other body parts, myc has been seen skulking about in a state of frightening disrepair. Sometimes it is torn into pieces and wildly rearranged, sometimes duplicated over and over again into aberrant circles of genetic excess, like little cysts of myc genes within the cell. The gene is so frequently mutated in cancer tissue, and in its normal guise it bears so many trademarks of being critical to the life and upkeep of all body cells, that some have called it McGene: everywhere you look, there it is, the myc gene.

After nearly two decades of alternately dabbling with the gene and then abandoning it as too hard to decipher, researchers have made discoveries that bring this extremely important molecule into focus. Many of the results are basic revelations about how a cell knows when to divide, when to mature, and, on occasion, when to commit suicide for the good of the body. The findings have implications for the treatment of cancer, most notably as a prognostic tool to distinguish between early breast tumors, which can be treated successfully through simple surgery, and highly aggressive cancers, which are likely to recur and thus should be treated with the most blistering chemotherapy drugs available. Studies from the Netherlands indicate that women whose breast cancer cells harbor an abnormally high number of myc genes, a type of mutation called gene amplification, are far more likely to suffer a recurrence of their disease after surgery than are those women whose malignancies lack signs of myc redundancy.

Test-tube and mouse experiments also suggest that of the many genetic flaws in the average cancer cell, the myc defect is so nasty that its elimination alone may be enough to cure or at least tame a substantial fraction of tumors. Using medications that have been available since the 1960s, scientists have managed to correct myc defects in cultured cancer cells. The drugs work by encouraging the cells to eject their excess copies of the gene, to boot them right out the door of the nucleus for speedy degradation within the cytoplasm; and once those amplified genes are eliminated, the cells calm down and revert to a noncancerous state. One of these drugs is now being tested for use in women with advanced ovarian cancer whose tumors display evidence of gene amplification.

As so often happens these days, however, our basic understanding of this McCancer gene far outstrips any clinical advances or applications. We now have a fairly good grasp of what the protein produced by the myc gene ordinarily does in the cell and why disrupting that housekeeping task has such disastrous consequences. The myc protein is a kind of toggle switch, sitting at the junction of two options an active cell must choose between: to proliferate with aerobic vigor, or to differentiate into a sedentary state as a mature member of the lung, colon, breast, or other organ. The myc protein, which seems to be necessary for a cell to begin dividing and to keep dividing, must be firmly silenced before a cell can mature into its final stage. Yet this proteinous master of cell proliferation is also a master of self-destruction. When a cell receives a shock, when messages bombarding its surface get crossed, or when it is momentarily deprived of nutrients, myc stanches the confusion by clicking on the button for self-destruction. It starts the violent chain reaction of apoptosis, culminating in the cell's demise.

The death is as dramatic as the death of a star or city. The cell's surface boils and ruffles, the DNA within topples like columns in an earthquake, the contents of the cell innards blow outward. Within twenty-five minutes, the cell is gone. While the idea of linking cell death to cell growth seems counterintuitive, evolution had good reason to make the connection. Cells must be able to divide in a healthy body to replenish lost tissue, but if that division encounters any sort of difficulty, any disturbance in surrounding biochemical signals, and the possibility of renegade growth arises, then the safest course for the cell is to set in motion a suicide program. After all, the most dangerous thing that can happen to a body is to have one of its cells turn malignant. You can have massive tissue necrosis — you can lose your limbs, three quarters of your liver, much of your brain — and still you'll survive, but one tiny cell with aspirations toward immortality can spell your end. How better to make certain that the potentially psychotic cell will commit suicide than to recruit the protein that normally helps ensure cell proliferation?

The renaissance in the myc field comes after years of stagnation in the analysis of the gene. Scientists first identified it while studying cancer of the bone marrow in chickens, and the gene won its name for being a promoter of *myc*loma in *c*hickens. Myc abnormalities were soon detected in many human cancers as well. Biologists were particularly excited when they discovered in the 1980s that some human leukemias and lymphomas result from chromosomal disasters called myc translocations.

At some point in the early stages of a malignancy, the part of one chromosome that holds the myc gene exchanges pieces with another chromosome, essentially shifting the gene from its rightful position and sticking it where it does not belong. This rearrangement frees the myc gene from the conventional chromosomal signals that keep it in check, and exposes it to other, more robust genetic switches, like the one that sets off the body's incessant production of immune antibodies. Relocated to its new position and turned to a permanently active position, the myc gene thenceforth can generate its protein nonstop, a mishap that fosters cancerous growth.

Other tumors were found to be the result of myc amplification, where the cancer cells are bloated with dozens or hundreds of extra copies of the gene, again resulting in the overproduction of the myc protein and the consequent lawless cell growth.

By the late 1980s, however, research into myc came to a standstill. Clearly the protein was terribly important, but it was so difficult to isolate that nobody could figure out what the molecule did or why it contributed to cancer when propagated in excess amounts. That state of despondency changed radically in 1990, when it was learned that the myc protein bore several telltale shapes, including one evocatively called a leucine zipper. The zipper motif had been seen on other proteins known to need a protein mate before they go to work in the cell. Parts of these proteins stick out like the teeth of a zipper, a design that makes the protein interdigitate with another protein that has bared leucine teeth. Once they are zipped together, the two proteins become a functioning unit, able to carry out a chore that either on its own can't manage.

By detecting vanishingly small amounts of protein and probing through tens of thousands of possible partners, biologists were able to root out the myc protein's mate. It was designated max, both because it acts on myc and because the two names, myc and max, sounded so catchy together. With the working myc and max complex firmly in test tube, scientists realized why the two proteins bothered zipping together to begin with. The complex proved to be a transcription factor, a protein bundle designed to pinch on to certain stretches of the DNA molecule and switch a series of genes on or off, as a way of initiating vast changes in the cell.

It's still unclear which genes the myc-max oligarchy commands to be active or be still. Hundreds if not thousands of the 100,000 genes in human DNA may possess the sequence that the complex recognizes, and it could take years to sort out how many of those genes the pair does communicate with, and how that dialogue results in cell growth, maturation, or, if need be, *seppuku*. However, at least one conversation partner of the pair has been identified. Myc and max appear to initate cell growth by shutting off the retinoblastoma protein. This protein normally serves to suppress the cleavage of one cell into two by blocking the activity of cell-division proteins. In other words, myc and max help start cell growth by silencing the actor that stops cell growth, just the sort of loop-the-loop, Rube Goldberg arrangement that nature seems to delight in.

39

THE OTHER SIDE

OF SUICIDE

CONSIDERED on its face, suicide flouts the laws of nature, slashing through the sturdy instinct that wills all beings to fight for their lives until they can fight no longer.

Yet by a coolheaded evolutionary accounting, suicide cannot be entirely explained as a violent aberration or a human pathology lying outside the ebbs and pulls of natural selection and adaptation. Suicide, for all its private, tangled sorrows, is surprisingly common in most countries, accounting on average for about 1 percent of all deaths. And when the number of unsuccessful suicide attempts is taken into account, the prevalence of the behavior jumps considerably. The incidence, some evolutionary geneticists say, is too great to be accounted for by standard explanations like social malaise or random cases of psychiatric disease.

Instead, the persistence of suicide at a high rate across most cultures of the world suggests an underlying evolutionary component, a possible Darwinian rationale for an act that too often appears starkly irrational. The inclination toward suicide could be a concomitant of a trait or group of traits that at some point in evolutionary history conferred benefits on those who bore it.

Further bolstering the case for a genetic basis to suicide is its tendency to run in families. Although suicide occurs in nearly all countries, it is far more common among some ethnic groups than others. The Hungarians and Finns, for example, suffer from suicide rates two to three times those in the United States and most of Europe. Significantly, the Hungarians and Finns are thought to share genetic roots in the distant past (as well as the linguistic roots that bind the Hungarian and Finnish tongues and set them apart from Indo-European languages). In addition, the elevated incidence of suicide holds true not only in those nations, where socioeconomic conditions could be responsible, but also for Finns and Hungarians who have emigrated to other countries, again hinting at a biological substrate.

Nobody argues that there is a single gene for suicide, or that suicide or mental illness should be thought of as good. The lure of suicide too often beckons to the young, who clothe it in the romantic chiffon of nobility and poetry and see it as a reasonable option should the transition to adulthood prove too traumatic, a way of thinking that no sane adult would condone.

Nevertheless, there may be plausible evolutionary explanations for at least some self-destructive acts. A number of theorists propose that the impulse to kill oneself may be an expression of an instinct toward self-sacrifice for the good of surviving relatives, either because those relatives will be rescued from their own death, or because they will benefit richly from the resources that will now accrue to them. The surviving relatives will in turn pass on the sacrificial victim's genes. To take a short and admittedly simplistic example, a hominid in the jungle may have enhanced his genetic survival by sacrificing himself to a leopard that would otherwise have slain six of his brothers or sisters. However, because we live in complex social groups, such an impulse toward martyrdom might on occasion show itself in complex, distorted forms, tugging miserably at the psyches of even those who have no families to benefit from their deaths or to ensure that their genetic legacies survive.

In another scenario, suicide is viewed not as heritable but as the most tragic outcome of another trait that may derive from natural selection — the tendency toward depression. Some Darwinian thinkers say that extremely bleak moods are themselves too common to be the result of pathology alone. They propose that bouts of depression may be useful, forcing people into a kind of emotional hibernation and giving them time to reflect on their mistakes. But such a strategy, if sustained too long or repeated too often, becomes maladaptive and even fatal, showing itself as the harrowing disease called major depression.

Reasoning that human beings invent few traits but instead display intricate versions of behaviors seen elsewhere in the animal kingdom, some biologists have looked to other species for clues to the genesis of suicide and depression. The exercise is fraught with perils. Nonhuman animals obviously do not leave behind anything as clear as a note, nor are they likely to have sufficient awareness to do something as deliberate as jump off a cliff. Still, there are numerous examples of creatures that sacrifice themselves for their kin, including termites that explode their guts, releasing the slimy, foul contents over enemies that threaten their nest, and rodents that deliberately starve themselves to death rather than risk spreading an infection to others in their burrow. What is more compelling, many species of nonhuman primates will suffer serious depression when stressed; on falling into an episode of melancholy, the monkeys may engage in all sorts of life-threatening activities — refusing food until they die of malnutrition, or swinging from dangerous tree limbs that no normal monkey would go near. So similar is monkey depression to our own that the symptoms of the mood disorder dissipate when the primates are given an antidepressant like Prozac.

Admittedly, one must approach this theoretical terrain with enormous trepidation, for it's all too easy to sound insensitive or glib in ascribing suicide and depression to the handiwork of natural selection. Psychiatrists have struggled long and hard to get the public to view mental illness as an organic disorder rather than a self-indulgence or character flaw, and most are reluctant to describe something like depression in anything other than the most strictly disease-oriented and condemnatory terms — as the mind's version of diabetes or cancer. Researchers know too well how easily a Darwinian explanation for complex behaviors can be overdone and oversimplified.

Certainly the affairs of animals much simpler than people have been misinterpreted in the past. For example, the mention of suicide in nonhuman species invariably raises the famed case of lemmings, rodents that were long thought to kill themselves en masse by running into the sea, as though alerted by a group alarm clock that today is a good day to die. As recent research has revealed, however, the tale of the suicidal lemming is false. The tawny, thickset rodents will die by the group, but that is a result of an error in judgment. Lemmings are the locusts of mammals, and they will strip a habitat bare. They then begin migrating to find new feeding grounds, swarming over boulders, around trees, whatever stands in their way. If they run into a body of water, they try to swim across, a routine that works fine for streams and ponds. If they happen to hit a lake or an ocean, they discover too late into their paddling that they can't make it.

Often it is not clear whether a death in the wilderness is deliberate or accidental. Some animal behaviorists have developed models predicting that, under certain circumstances, a hatchling bird in a multichick brood does better from the perspective of its genetic legacy to let itself be killed by its siblings than to fight back. Among crested penguins, for instance, a mother always lays two eggs a season, one large, one small. Given the harshness of her arctic surroundings, she can rear only one bird to independence, and usually that lucky penguin will come from the large egg. Still, she lays the double dose as an insurance policy, in case the big egg is preyed on. Should the big and little eggs both end up hatching, the smaller offspring in theory would do best to permit the bigger sibling to kill it off without putting up a fuss — essentially, to throw itself on its sibling's sword. After all, both birds can't possibly survive, so why divert resources from the relative with the greater chance of success?

The theory has some observational data to back it up: in

encounters between sibling penguins, the little ones do appear to die off without ruffling anybody's feathers. However, critics of the scenario do not buy that the smaller contender is going gently into that good night. They point out that if your ordinary Joe were stuck in a lifeboat with Mike Tyson and a very limited quantity of food, the nonboxer would be foolish to challenge Tyson to a fight; instead, the little fellow is likely to lie low and look for a chance to push him overboard, or simply pray that Tyson gets struck by lightning.

In general, scientists will call a death a suicide only when the animal has much to gain and little to lose, reproductively speaking, from the act. Among this group some put cryptically colored butterflies that escape being eaten by blending into their surroundings. Once an adult is past its reproductive time, it becomes a risk to its surviving offspring, for if the elder insect is discovered by a bird, the predators will gain clues from the butterfly's pattern to discriminate prey from background; thereafter, the younger butterflies will also be in danger. As it happens, the postfertile adults are known to drop to the groundcover and begin beating their wings rapidly until they die of exhaustion. They obliterate themselves and their secrets before they get caught.

Other exemplars of self-sacrifice abound. In some gall midges — tiny gnatlike insects — a mother offers up her body as a meal to offspring, and they happily consume every last segmented bit of her. Among naked mole rats, hairless and blind rodents that live underground and are almost as closely related to one another as are bees in a hive, an animal that is infested with parasites knows what it must do: go off to the communal toilet area of the burrow and remain there until it dies. Once its decision is made, it won't move, and it can't be force-fed, even under laboratory conditions; the sickly mole rat will not risk infecting the whole colony.

In applying to human beings the idea of self-sacrifice for the good of one's kith and kin, scientists cite obvious examples: mothers who gladly die to save their children, war heroes who go down in flames for their buddies, or even the recent spate of so-called rational suicides, in which elderly or terminally ill patients request that they be allowed to die quickly to avoid being a drain on their families. Researchers are much more reluctant to use such reasoning to justify the behavior of suicidal patients who very often are mentally ill, lonely, and alienated from those who care about them. Yet psychiatrists observe that people who are contemplating suicide frequently think of it in extravagantly selfless terms, as the option that would be best for the suicide's family and friends. Those who have talked to people immediately after they made a serious suicide attempt report that the patients often have an altruistic explanation for what they did, believing the action to be the wise, clever, and thoughtful thing to do. In that sense, our own version of suicide sounds remarkably similar to the response of a parasitized naked mole rat. Those who consider killing themselves feel grotesque, polluted, infected, and they may think it best to destroy the source of disease before it contaminates their loved ones.

Of course, it's true that the majority of those who attempt suicide are afflicted by a disease — a mood disorder, in most cases depression or manic-depression. Such conditions are characterized by a dramatic drop in neurotransmitters like norepinephrine or serotonin, the molecules that allow nerve cells to communicate and that help modulate emotions and aggression.

Studying nonhuman primates, scientists say they have witnessed many of the symptoms of serious depression in their subjects, including a disregard for self-preservation. Jane Goodall, the renowned chimpanzee champion, once observed a seven-and-a-half-year-old male chimp experience such profound despair after the death of his mother that he refused to leave her corpse even to eat. The monkey slowly withered away, lay down, and died — of a broken heart, Ms. Goodall said.

Monkey depression resembles our own not only behaviorally but biochemically. Working with a free-ranging colony of rhesus monkeys, behavioral scientists have learned that about 20 percent of the primates are predisposed to serious depression, roughly the same as our own lifetime risk of the disorder. The monkeys fall into their slump when they lose a relative or close partner, or suffer a drop in social status, events that likewise set off human depression. In addition, the depressed monkeys show some of the same changes in brain chemistry that have been observed in human patients, including a drop in cerebrospinal levels of norepinephrine.

Depression, then, is evolutionarily ancient, preceding the appearance of hominids. Its purpose could be protective, allowing people and other animals with advanced cognitive skills to assess their situations, consider how their tactics may have backfired, and figure out a way to avoid repetitions of the costly error. Alternatively, depression could be the inevitable downside of a personality that offers great payoffs when times are good. In the case of the rhesus monkey, the same animals that are susceptible to depression often rise to the top of the social hierarchy because of their heightened sensory and emotional sensitivity. They're more aware than their peers of critical changes in the environment — the sound of an approaching predator, the tentative gestures of a possible new ally. They see and hear and smell everything. They are like taut strings on a beautiful violin. If bowed too hard, the strings will snap.

40

ANOTHER STITCH
IN THE QUILT

I DON'T BELIEVE there is such a thing as a good way to die. Death to me is a wasteful obscenity. You spend your life mastering tasks, cultivating knowledge and opinions, gradually getting the hang of living in your skin and skull, when it all must be disposed of to make room for the latest models coming up from behind. Nature is a spoiled brat who needs a perpetual supply of new toys. So how can you die well when death is such an imposition?

But if there's anybody I know who died with style — a style entirely appropriate to his fidgety, careless, extravagant personality — it was Rodney Holmes, a friend of my family's from the time I was four. Rod died quickly and catastrophically, within a few days of contracting a pneumonia so severe that it sent him into a coma. The doctors tried shooting him up with every antibiotic on the shelf, but he never regained consciousness. He didn't linger, he didn't waste away, he didn't pay attention. Not long before his death, I'd met him at a health club, and he'd looked fit and hale. But when he entered the hospital, his T4 cell count was 25, against the normal reading of 1000 or so. Rod died of complications from AIDS, and he had never even known that he was infected with the human immunodeficiency virus. He didn't know because he didn't want to know.

I suspect there were two reasons that Rod chose to remain ignorant of his HIV status. One is that he knew himself. He knew he wasn't a fighter or a confronter or a take-charge, stand-up-and-shout sort of man. He wasn't the type to scour the medical literature and familiarize himself with every new drug that offered a thread of hope. Instead, he had a tendency — one that I know only too well — to manage problems by ignoring them. If trouble comes to greet you, slam the door in its face; that at least gives you a little breathing room while trouble is figuring out how to jimmy the lock. Fighting an illness well probably requires a meticulous, organized approach to life, and Rod had anything but. He was a slapdasher and a barnstormer. He had brilliance as an interior designer and carpenter; but when he was working on a project, he tried to get it done so fast that he left corners unmatched, edges unpainted, nails unhammered. Fighting an illness also demands a certain degree of self-confidence, and Rod flaunted his insecurities raucously, aggressively. He was a short, unhandsome man with a face that was a cross between a bulldog's and a porcelain doll's; and I'll bet it's even harder to be a homely gay man than it is to be a plain woman. His wit seared and his imagination soared, and he worked mightily to entertain those he was with; yet he had few close friends, and those he had were mindful of his flaws, his impossible, incomparable nature. Even at his wake, the pastor, who'd known Rod well, lapsed into a spell of funeral verité, spending as much time recounting Rod's vices as he did his virtues.

Rod had succeeded in wrestling one immense problem to the ground: his alcoholism. He'd been an outrageous drunk for much of his adulthood, but then he joined Alcoholics Anonymous and stayed sober for the last eight years of his life. Sometimes I think he felt he had to choose between sobriety and self-awareness. It was only shortly after he gave up drinking, in the mid-1980s, that he came down with the first possible sign of HIV infection — a case of shingles. But even though I and others urged him to get tested for the virus, he refused to discuss it. Maybe a positive result would have been too much for him to bear without a fuzzy blanket of booze. Or maybe at some level he was trying to protect others, knowing that he would be much likelier to go out and spread the virus around were he to lapse back into drunkenness. Or maybe he realized, in a moment of plainspeaking terror, that AIDS was much, much bigger than he or his will or his wit or his vision. AIDS is bigger than anything any of us has ever seen. In its virulence, its absolutism, the way it has singled out and eviscerated our artists, our dissidents, our source of the creative and nourishing new, AIDS ranks as one of the most vicious diseases the human race has ever seen.

Part of the ugliness is the virus's method, the completeness with which it tears the body to bits. It works so well because it has learned what no human microbe had ever learned: to systematically target the host's defenses against it. The virus infects and kills helper T cells, the generals of the entire immune army; these are the T4 cells whose count among the poz-cognoscenti is shorthand for one's prospects. It kills killer T cells; it kills macrophages; it utterly upends the body's network of cytokines, the signaling molecules that allow immune cells to communicate. HIV mutates restlessly: the moment its face grows familiar to the antibodies seeking to neutralize it, it dons a new proteinous mask and slips past unheeded. Even when it seems to be latent, even when you can see no trace of it in the bloodstream, the virus is quietly disassembling the immune system. For years, it propagates in the lymph nodes, where it ambushes the T cells that invariably wander through on their transcorporeal journeys.

In destroying the immune system, the virus destroys the scaffolding of the self. Immune cells set the boundary between self and others. They distinguish flesh of one's flesh from the squatter, the bloodsucker, the thief. Unless there is an immune system, the gates can be flung wide, allowing all to storm the sanctum. Secondary infections that normally would be waved off like so many gnats instead burrow in and proliferate into choking swarms of fungi, bacteria, viruses, protozoa, yeast. More than any other disease, AIDS reminds us that we are organic matter, a rich natural resource, a meal for the invisible multitudes. The scavenging usually awaits our death; a patient with AIDS is eaten alive.

AIDS has butchered many people, of course. As of the end of 1994, nearly 250,000 people had died of the disease in the United States alone. But the misery extends farther than the stack of death certificates suggests. AIDS has also gutted our courage so thoroughly that we hardly notice it's gone. AIDS has twisted and turvied the most intimate expressions of love into something approaching homicidal or suicidal acts. It has been a carnival for the prigs and scolds and bullies among us, those who have always seen eroticism and sexual adventurism as a menace, and now claim to have history and science on their side. The conservative revolution may have begun before the outbreak — Ronald Reagan was elected in 1980, a year before the first reports of the disease became public — but AIDS has stoked the flames of right-wing fury and millennial brimstone. After all, the disease has remained largely confined to such high-risk groups as gay men and intravenous drug users, just the sorts of people who had already been demonized and marginalized. In defiance of early predictions, mainstream heterosexual America has largely been spared from the holocaust, which bolsters the opinions of bigots that God is white and straight and probably Christian, too.

In the age of AIDS, a dreary neo-Victorianism is everywhere in evidence. Nobody today will defend promiscuity and sexual passion as forms of self-expression or self-discovery, and even the most stylistically outré artist is likely to be a sexual prude. Virginity before marriage is back in vogue, sentimentally if not literally. People who have tried to find silver linings in this blackest of all storms say that the threat of AIDS has forced them to take themselves more seriously, and to seek deeper and more meaningful ties with their partners. That's a fine thing to say if it makes you feel better, but I don't buy it. There are many magnificent reasons to be faithful to the one you love; fear isn't among them. If anything, fear is likelier than familiarity to breed contempt.

I feel nothing but revulsion toward AIDS, a disease that in this country perversely culls the young, the vivid, the joies of our vivre. AIDS has throttled hope and baffled medical science, destroying the brief illusion that the age of infectious diseases was behind us. Nor has the tragedy brought us together and inspired us to new heights of generosity, insight, and empathy. If anything, AIDS has strengthened old partitions between people and constructed a few of its own: between straights and gays, between the poor and the well-to-do, between the developing nations, where the rate of new HIV infections is galloping out of control, and the developed world, where the number of new cases has leveled off. Even among male homosexuals, HIV serves as a rampart, a perverse maker of tribes, separating those who are positive — and coping and worrying and wondering on each new date when to reveal their viral status — from those who are negative and want to keep it that way.

Rod wanted nothing to do with this new form of social clubbery, this novel method of discrimination within the subculture of which he was already an uneasy part. He could be a terrible snob, mocking those he thought lacked his aesthetic sense, his perceptiveness, his venom-tipped wit. But that was just a plain old human swill of vanity and insecurity, not a life-and-death discrimination or a biblical separation between the damned and the saved. With his alcoholism, his funny looks, his thistly personality, his unstable income, Rod spent his life on the fringe, on the out, set apart, and he didn't need yet another stigma to make him lonely. To the extent that a gay New York man can ignore the AIDS epidemic, Rod did. He didn't have many gay friends, so he didn't go to funerals. When he asked his young lover to move in with him, the man's medical history — his infection with HIV and his tuberculosis — was of no concern to Rod. He was madly, obsessively in love, and that alone was what he wanted to talk about.

At the time, I thought Rod was foolish for risking his health so heedlessly, but as I reconstruct events, I figure he was already infected by the time the man entered his life. Maybe he suspected as much, but if so he didn't act on it — no AZT, no ddI, no aerosolized pentamadine to ward off pneumonia. As a result of his refusal to acknowledge AIDS or take any medicine that might have slowed its course, Rod died in 1993 the way victims did at the beginning of the epidemic, before doctors even knew what they were seeing. He wasn't laid waste piece by piece. AIDS just came, and swallowed him whole.

41

A GRANDDAUGHTER'S FEAR

I WAS TALKING business with a colleague late one afternoon when the first call came. The voice on the other end was so loud and panicked that my colleague could hear it and looked over at me, horrified. "Natalie, is that you?" screamed my grandmother. "Natalie, I need you to do me a favor!" I cupped the telephone mouthpiece. "What's wrong, Grandma?" She started to cry, her voice heaving and gasping. She told me that she'd been alone since twelve o'clock — for five hours! — and that nobody was scheduled to visit her until ten that night. She couldn't stand it; she was going crazy; they'd left her all alone; and she was so afraid. She'd been calling everybody, everybody — her son (my uncle), her stepson, her grandsons. And now me. "So what do you want me to do?" I muttered, although I knew the answer. "Please, darling, come over! Can't you come here? Please, Natalie! I'm ALL ALONE!"

"OK, OK," I said. "I'm coming over. I'll be there as soon as I can."

I hung up and told my colleague that I'd have to leave in a few minutes. But I didn't think there was any immediate crisis. My grandmother lives in an apartment building on New York's Upper East Side where plenty of people know her and stop by to see how she is. She just wanted company — and, damn it, I was busy. I continued my business discussion for another twenty minutes. Until the second call came. This time my grandmother was genuinely hysterical. "Natalie, WHERE ARE YOU?" she cried. "You said you were coming right over! Please, darling!"

Now I did hurry, racing out the door and grabbing a cab. But the moment I arrived at her apartment I wanted to run the other way. She clutched my arm and pulled me inside. Her face was a gray blur of tears and her thin white hair stood up in wild peaks. Her bathrobe had half fallen off, and I turned away, embarrassed; I'd never seen my grandmother's naked body before. The apartment smelled stale, suffocating. I threw open a window and sat down stiffly on the couch.

For the next few minutes I didn't say a word while my grandmother shuffled around the living room, ranting against the world. She complained about my uncle, who had deserted her earlier in the day. (He had to go to work.) She railed against my mother — her daughter — who was vacationing in Australia. She talked madly about how they were trying to steal her money, how they'd taken away her keys, how they never spent more than ten minutes at a time with her — although I knew only too well that both of her children structured their days and nights around her needs.

As she sputtered on, I felt more and more helpless and resentful. Finally, my rage overwhelmed my judgment. I stood up and started yelling at her. I told her that nobody and nothing could help her. "The only person who can help you is you!" I said in righteous fury. "Do you understand me? You've got to stop being so damned dependent on everybody!" At which point she let out a piercing shriek and hurled herself on her bed. The only thing I'd accomplished with my idiotic lecture was to heighten both her hysteria and my sense of impotence.

My grandmother is eighty, but she seems much older. Although she suffers from a host of physical ailments — mild diabetes, glaucoma, asthma, arthritis — her real problems are neurological and psychological. She may have Alzheimer's; she may have been stricken by a series of small, silent strokes. Her doctor isn't sure, and he says that, frankly, the precise diagnosis doesn't matter: her condition is irreversible. What is clear is that she hates being old, she can't stand being left alone for even minutes at a time, and she'll do anything to surround herself with people, particularly relatives who, being relatives, owe her their lives.

These days, much of the family conversation centers on her. What are we going to do about Grandma? Put her in a nursing home? (Too awful.) Hire a live-in companion? (Too expensive.) Put her on some new psychotherapeutic medication? (Nothing seems to work.) Not only does my grandmother demand companionship during the day; she needs somebody around every night. So another question my family grapples with is whose turn it is to sleep over on Grandma's sofa bed. By far the most debilitating consequence of the ordeal is the guilt. Because we can't seem to make my grandmother happy, we feel frustrated. The frustration leads us either to explode in anger or to drop out of sight — immature reactions that come packaged with shame. No matter how much she does, my mother worries that she's not doing enough. At the same time, she bitterly resents her mother's nonstop demands, threaded through as they are with insults and accusations. The result is that my mother visits and calls my grandmother constantly, but ends up lashing out at her in senseless indignation. My uncle usually represses his emotions, but he's starting to gain weight and to look his fifty-six years.

I manage to combine the worst of all worlds. I neglect to call my grandmother for weeks at a stretch. When I do visit, I lapse into the role of boot-camp sergeant. As a fitness fanatic, I tell her it's never too late to take up exercise. I turn away from her tears. I pick up a book and start reading while she's in the middle of a mad monologue. My mother accuses me of being heartless, and she's right.

My grandmother is the first person I have watched grow old. I used to adore her; she still keeps loving poems and letters I wrote to her as a child. She was always a vivid, energetic woman, selling bonds for Israel, working long hours in charity thrift shops. She told stories about her past with the narrative panache of Isaac Bashevis Singer. Wherever she went, she made flocks of friends — a trait that I, a lonely and sullen girl, profoundly admired.

But then hard times began to pile up around her like layers of choking silt. Although she'd stoutly nursed three husbands through terminal illnesses, she became depressed when her siblings — all older than she — started to die. After she lost her last sister, in 1982, my grandmother just about lost her mind. She still had many friends, but she clamored for ever more attention from her children and grandchildren. She became an emotional hair-trigger; she had temper tantrums at parties, seders, my sister's wedding.

As my grandmother has worsened, so has my response to her. My mother implores me to be decent and stay in touch, and I launch into all the reasons that I don't. But my excuses sound shallow and glib, even to me. The truth is that my grandmother terrifies me.

I have in my mind a pastel confection of the perfect old woman. She is wise and dignified, at peace with herself, and quietly proud of the life she has forged. She doesn't waste time seeking approval or cursing the galaxy. Instead, she works at her craft. She is Georgia O'Keeffe painting, Louise Nevelson sculpturing, Marianne Moore writing. Or she is a less celebrated woman, who reads, listens to Bach, and threads together the scattered days into a private whole.

Of course, there are many things my fantasy doyenne is not. She's not strapped for money. Her joints don't ache, and her breath doesn't rattle. She isn't losing her memory, her eyesight, her reason. Above all, she is not the old woman I know best.

I love my grandmother. She still has her good hours, when her mind is quick and clear. Inevitably, though, her mad despair bursts to the surface again. She discovers a new reason to weep, blame, and backstab, and I discover a new excuse for staying away.

I want to age magnificently, as O'Keeffe and Moore did. I

want to be better in half a century than I am at thirty-one, but I doubt that I will be. When I look at my grandmother, fragile, frightened, unhappy, wanting to die but clinging desperately to life, I see myself — and I cannot stand the sight.

My grandmother died in September of 1991, a day before her eighty-third birthday. I still dream about her, more often now, in fact, than I did immediately after her death. In my dreams she is always much younger than she was when she died, and she is always completely sane and strong. She becomes herself again, my grandmother at her best, and I stare at her in wonderment and relief. My dreaming mind is a child's mind — sentimental, grandiose, rewriting stories so that their endings can be born.

INDEX

adenine, 92
adipose tissue, 189, 190. *See also* fat
adrenaline, 194; and fat cells, 192
African sleeping sickness, 107
aging: author's grandmother, 256–60; degenerative diseases of, 55; telomeres and, 72–73, 75–76
AIDS, 250–55; and DNA bending, 79; King's research on, 221; number of deaths from, 253
alcoholism: and creativity, 206
Alzheimer's disease, 55, 234; ubiquitin and, 237
American cockroach, 120
amidinohydrazones, 116–17
androgens: in hyenas, 143; and mammalian development, 139, 140
androstenedione: in female

primates, 143; in hyenas, 142, 143
Angraecum sesquipedale, 46
animal behaviorists, 23, 171–74
animal development: Hox genes and, 83–88
anteaters, 133
anthropomorphism in science, xiii, xiv–xv, 169–74; advocates for, 169–71, 174; traditional criticism of, 169, 171, 173–74
anticarcinogens: dietary, 182–87
antioxidant enzymes, 185
ants: activity of, 165; parasites on, 108
apoptosis, 233–37, 240; discovery of, 235
arachnids, 97–102
Argentina's "dirty war," 219, 222–23
Aristotle, 205